FUNDAMENTALS
OF
OROFACIAL MYOLOGY

FUNDAMENTALS OF OROFACIAL MYOLOGY

By

MARVIN L. HANSON, PH.D.

Speech-Language Pathologist
and
Orofacial Myologist
Chairman, Department of
Communication Disorders
University of Utah
Salt Lake City, Utah

and

RICHARD H. BARRETT, B.A., M.ED.

Speech-Language Pathologist
and
Orofacial Myologist
Berryville, Arkansas

C H A R L E S C T H O M A S · P U B L I S H E R
Springfield · Illinois · U.S.A.

72284

SEP 0 7. 1989

Published and Distributed Throughout the World by

CHARLES C THOMAS • PUBLISHER
2600 South First Street
Springfield, Illinois 62794-9265

© *1988 by* CHARLES C THOMAS • PUBLISHER

ISBN 0-398-05518-1

Library of Congress Catalog Card Number: 88-20041

With THOMAS BOOKS *careful attention is given to all details of manufacturing
and design. It is the Publisher's desire to present books that are satisfactory as to their
physical qualities and artistic possibilities and appropriate for their particular use.*
THOMAS BOOKS *will be true to those laws of quality that assure a good name
and good will.*

Printed in the United States of America
SC-R-3

Library of Congress Cataloging-in-Publication Data
Hanson, Marvin L., 1932–
    Fundamentals of orofacial myology / by Marvin L. Hanson and
Richard H. Barrett.
      p. cm.
    Includes bibliographies and indexes.
    ISBN 0-398-05518-1
    1. Tongue thrust. 2. Oral habits. 3. Orthodontics. I. Barrett,
Richard H. (Richard Howard), 1915–   . II. Title.
    [DNLM: 1. Fingersucking—therapy. 2. Mouth Rehabilitation.
3. Tongue Habits—therapy.   WU 140 H251f]
RC429.H36   1988
617'.522—dc19
DNLM/DLC
for Library of Congress   88-20041
                                              CIP

PREFACE

We hate to say we told you so, but. . . .

Our faith in the basic validity of the profession of orofacial myology is of long standing. It was established before there even *was* such a profession, and certainly before it was called orofacial myology. Our faith held steadfast during the 1960s, when the field was largely either scorned or ignored. It did not falter during the 1970s, when it seemed that controversies would never be resolved.

It is therefore gratifying to note the increased recognition and acceptance of this province in the late 1980s, and to anticipate still broader growth in the 1990s. We do not expect the profession to fully mature until the next century; by then, we foresee the field being taught on the college level as a distinct discipline.

Our last volume devoted to this subject was published ten years ago. We had expected the arrival of other contributions during this interval, but no comparable work has appeared. This book, which modifies and modernizes our previous editions, fills the resultant void.

Although moving in that direction, orofacial myology still has not developed its own body of controlled research and derived therefrom a unique system of knowledge. To a large degree, it remains a melange, an admixture compounded of elements borrowed from the professions it serves. Readers already trained in one or another of those professions may therefore consider some portions of the chapters that follow to be superfluous. Even so, they may be interested in observing how those concepts are synthesized into the total fabric comprising orofacial myology.

This is the one area where dentistry and speech/language pathology overlap, bringing about shared goals and concerns. The dental hygienist also becomes involved and, to a lesser extent, one or two other professionals. Each has only partial knowledge of the total field of orofacial myology.

The purpose of this book is to provide a complete guide to the area. We have attempted to extract from each component realm the requisite teachings that it can supply. This book presents a balance of theoretical and practical information thus gleaned, designed to supplement the training and knowledge of each basic field. It is therefore appropriate to serve as a textbook on the subject for advanced undergraduates and graduate students in dentistry, dental hygiene and speech/language pathology. All of this is in addition to the book's role as a text for direct training to becoming an orofacial myologist.

As we noted in the Preface to the first of our volumes, the learning of all the material included herein does not prepare the student to administer therapy. Book reading must be supplemented by extensive supervised work with patients of all ages, representing all of the disorders the field encompasses. We are now able to offer far more information than contained in our first edition, but there is still no substitute for guided experience.

M.L.H.
R.H.B.

CONTENTS

FUNDAMENTALS
OF
OROFACIAL MYOLOGY

CHAPTER ONE

TONGUE THRUST AND RELATED DISORDERS: AN OVERVIEW

Introduction

O rofacial myologists, speech-language pathologists, and orthodontists share a challenging problem: How to ensure permanence of
results. Tongue thrusts, stuttering, and crooked teeth can all be modified,
and the improvement substantiated with numbers. But what happens to
the tongue thrust, the stuttering, and the malocclusion after treatment is
terminated?

Few studies have been done in any of these areas. Three such studies
are pertinent to our considerations as orofacial myologists. In chronological order:

1. In 1981, Uhde reported finding unacceptable occlusion in 49.2
percent of 72 orthodontically treated patients examined twelve years
after the completion of treatment. Examination of these patients immediately following removal of retainers would probably have found acceptable occlusion in all of them.

2. Lopez-Gavito and associates (1985) found that more than 35 percent
of a group of 41 subjects seen an average of nine years six months out of
retention demonstrated a postretention open bite of 3 mm.

3. In 1987 Andrianopoulos and Hanson found significantly less orthodontic relapse in 17 patients seen an average of 7.4 years postretention
who had received therapy for tongue thrust, than in 17 patients a like
number of years following treatment who had not received therapy for
tongue thrust. All patients had Class II, Division I malocclusions prior
to orthodontic treatment. Those with therapy for tongue thrust had a
mean relapse of 0.59 mm; those who had had no therapy for tongue
thrust relapsed an average of 1.94 mm.

The editor of one reputable orthodontic journal rejected the Andrianopoulos-Hanson article, lauding the research it reported and the
quality of the writing, but stating that its topic was not clinically relevant.

The reason therapy for tongue thrust came into existence was that orthodontists were concerned about the considerable relapse they were finding, weeks, months, and years posttreatment, in their patients who demonstrated tongue thrusting behaviors. People who provide therapy for tongue thrust see more patients who have overjets than those with other malocclusions. That is because orthodontists treat more patients with overjets than those with open bites, deep overbites, protruding mandibles, crossbites, and other types of malocclusions. Approximately 50 percent to 70 percent of patients with overjets have tongue thrusts. Ninety-eight percent of patients with open bites have tongue thrusts. The problem of relapse following orthodontic treatment *is* of grave clinical significance. This chapter will review relationships among tongue thrust, malocclusions, and treatment of tongue thrust. It will present strong evidence that treatment for tongue thrust is effective, not only in the short run, but for years following the completion of that therapy.

Semantics. The title of this volume uses the currently accepted name for the study of oral habits and their treatment.

Names are important. People who work with tongue thrust and related habit disorders now call themselves orofacial myologists or oral myofunctional therapists. The field they work in is orofacial myology. It used to be called tongue thrust therapy, but that name was too limited in scope. The therapist must not only eliminate the habit of pushing the tongue against the front and side teeth, but must also normalize functions of lip and facial muscles.

The area of cleft palate is now called "Craniofacial Anomalies." That is a good name, because it is precise and covers the scope of disorders related to cleft lip and palate. It is not so easy to say as "cleft palate," but it is better. A better name for "orofacial myology" might be "Oral Craniofacial Disorders." Professionals who treat these disorders are concerned with the inside of the mouth, with interrelationships between and within mandibular and maxillary structures, with function of the facial muscles, and with the appearance of the face.

"Tongue thrust" has been given many names in the past. Perverted swallow, deviate swallow, deviate deglutition, reverse swallow, infantile swallow, and visceral swallow are a few. The one that survived, tongue thrust, is too limited in scope, but it conjures up memorable visual images, and gets at the most harmful part of the disordered patterns. The others have been pretty much discarded, except by those who initially coined them. We will take the liberty to use several terms in this volume, just to keep it interesting for you.

We are concerned with any habits, or behavioral patterns, that call attention to themselves because of their inappropriateness or that have an undesirable effect on teeth or on speech. These include thumb or finger sucking, tongue sucking, lip biting, object sucking or biting, or lip licking. The most harmful of all habits seems to be that of resting or pushing the tongue against the front teeth. Among speakers of English, the only time the tongue should have contact with the front teeth is while saying the "th" sound. It should rest against either the upper or lower gums, just behind the front teeth. It should squeeze up against the roof of the mouth during swallowing. When it pushes against the front or side teeth, or protrudes between the upper and lower teeth, a tongue thrust is occurring.

Tongue Thrust: Definition

That is a general definition of a tongue thrust. A more precise one is:

Habitual resting or pushing of the tongue against at least ½ of the lingual surface area of the incisors or cuspids, or protrusion between the upper and lower anterior teeth.

The stipulation that the tongue contact at least ½ the surface area is arbitrary, but if only the portion of the teeth near the gingiva is contacted by the tongue, it is doubtful that the teeth will be moved by the resulting force.

Descriptions: Normal and Abnormal Behaviors

Normal Behaviors

In order to understand, identify, and treat abnormal eating and swallowing patterns, a knowledge of *normal* patterns is essential. A traditional description is offered: A person takes a moderate-sized bite of food and chews with the lips closed, allowing cheek and lip muscles to move the food toward the tongue as s/he chews. S/he chews just long enough to allow the saliva to mix with the food and form a cohesive bolus, right in the middle of the upper surface of the tongue. The tongue tip is positioned against the upper alveolar process, the sides of the tongue are positioned against the gums along the sides of the arch, and no food is allowed to escape laterally or anteriorly during the swallow. The molars are occluded, the lip and cheek muscles are relaxed, and the food is moved posteriorly by a lifting or squeezing action of the tongue. First the blade lifts, and then the posterior portion lifts, while the tip and sides of the tongue retain their contact with the alveolar process. When the swallow is completed, the teeth and tongue are free of food particles. In 1813, Magendie described eating as consisting of three stages: oral, pharyngeal, and esophageal. Several sources have stated that the oral stage is both conscious and voluntary, the pharyngeal stage conscious but involuntary, and the esophageal stage both unconscious and involuntary. Having noted this tidy arrangement, some have then ignored its implications and proceeded as if the entire process were an unconscious, involuntary, global reflex.

Swallowing as a Reflex. Some examination of the description of swallowing as a "complex reflex" activity is thus necessary. Certainly it is a complex act, and there can be no dispute that portions of it are purely reflex. However, this description does not apply to all aspects of deglutition.

A reflex may be defined as the involuntary muscular contraction that results from the stimulation of a sense organ (Best and Taylor, 1961). With excitation of a receptor organ, a chain of events is set in motion that must be carried on to its irrevocable conclusion. It has been noted that swallowing is probably the most complex all-or-nothing reflex obtainable by peripheral nerve stimulation. But where does this reflex begin? Where are the end organs located that fire off the reflex? Most studies place them primarily in the tonsils and in the anterior and posterior pillars of the fauces, with other concentrations in the base of the tongue, the soft palate, and the posterior pharyngeal wall. A bolus reaches this region, and thus triggers the reflex, only at the conclusion of the oral stage of swallowing. To state that the second and third stages are reflex is quite correct, but if one wishes to initiate the reflex in the absence of food, it is first necessary to voluntarily collect saliva and voluntarily proceed through the oral stage of swallowing, after which the saliva may serve as a mechanical stimulus for the reflexive remainder. There is general agreement among authorities that the oral stage is not bound in the reflex; it is voluntary, and although it is usually unconscious, it may easily be called up to consciousness. It is doubtless performed in a habitual manner, but habit is quite different in terms of modification from reflex.

Even were the action a reflex, it could still be changed. It is only necessary to alter one element in the reflex arc to change the response. It may be noted also that the pupillary reaction to light, Babinski's sign, and the contraction of blood vessels are highly inaccessible reflexes on the whole; yet, using modern biofeedback procedures, or under even a relatively moderate level of hypnosis, they become accessible to alteration.

Oral Stage. This is the *only* stage of eating with which the clinician need be directly concerned. Any abnormality that occurs is present only in this stage; once the bolus is delivered to the oropharynx, it may be consigned to its ultimate destination with a light heart.

Mastication. It is well to have some picture of the oral action immediately preceding swallowing. Mastication is a complex activity in itself: it is voluntary, although not always conscious. Placing food in the mouth does not trigger a reflex, although, once initiated, it may be continued on a subcortical level. It is centrally regulated by a relatively large area in the inferior medial portion of the motor cortex. Mastication may also be considered in three stages: incision, crushing, and grinding.

Incision. Incision begins with a lowering and protrusion of the mandible to bring the incisal edges of the upper and lower teeth into functional relationship. During incisal penetration of the food, the mandible is elevated continuously so that the incisal edges of the lower teeth contact the uppers and pass on over the lingual surfaces of the upper teeth.

Crushing. The food thus ingested is placed by the tongue and cheek muscles between the occlusal surfaces of one side or the other to be crushed by the molars and bicuspids. The lips are routinely closed and all the facial muscles are subjected to strenuous exercise during forceful mastication. Crushing is accomplished by a simple hingelike lifting and lowering of the mandible.

Grinding. The mandible moves in a somewhat rotary fashion, with the bolus being shifted occasionally from one side of the dental arch to the other. There is negligible tooth-to-tooth contact during mastication, the stroke being reversed immediately at the first proprioceptive signal of impending contact.

Bolus Formation. The spatulating action of the tongue maintains the food in a fairly cohesive bolus while mixing in the mucous and salivary secretions of the sublingual and submaxillary glands, thus serving to moisten and lubricate the reduced particles. Contractions of the buccinator muscle impregnate the bolus with ptyalin and serous fluid secreted by the parotid gland. The bolus is then centered on the dorsum of the tongue. The manner in which this is done is of vital importance to the clinician, for it is a pivotal concept in the retraining program presented herein. The lips and cheeks suck against the outer surfaces of the teeth, pulling the bolus into a reasonably cohesive unit on the tongue (Strang and Thompson, 1958). This sucking action is a critical factor in initiating normal deglutition; it not only positions the bolus but to some degree continues through the entire oral stage. Facial muscles are active in the collection of dispersed food particles, or of saliva in the absence of food.

Bolus Movement. The first discrete movement preparatory to swallowing is a depression of the apex of the tongue as the bolus is moved forward in the mouth. A sucking action centers the bolus in a groove on the dorsum of the tongue; with a combined sucking and lifting action the tongue is then raised and a seal is established between the periphery of the tongue and the hard palate. The tip of the tongue at this point is most commonly on, or slightly anterior to, the incisal papilla, in which position it is able to achieve some stability; this is the free portion of the tongue (the muscles have no skeletal attachment) and to function efficiently they must therefore seek "anchorage" through sheer pressure against the alveolar ridge. The ingesta are completely circumscribed at this moment. The lateral margins of the tongue seal against the buccal teeth and adjacent palatal mucosa. Posteriorly, the pharyngeal portion of the tongue arches behind the bolus. The posterior pillars contract toward the midline, and the depressed soft palate moves inferiorly to seal against the tongue (Wildman et al., 1964).

Practically all the intrinsic and extrinsic muscles of the tongue, plus the suprahyoid muscles, are active as the bolus is positioned and propelled. In addition, the muscles of mastication routinely hold the teeth in firm occlusion, thereby supporting the act with increased mechanical stability, particularly with coarse food or a large bolus. However, molar occlusion is not essential to normal deglutition, for once the bolus is trapped between tongue and palate, there is less concern for what occurs in the oral cavity below the level of the seal.

Then begins the phase that has immortalized Ardran and Kemp (1955) for in almost every description of swallowing is an echo of their analogy of toothpaste being squeezed from a tube. The tongue presses forcefully upward in a wave of distal motion that has been characterized as a "stripping wave." The apex and lateral aspects of the tongue remain fixed, preventing escape of the bolus, while the pharyngeal segment of the tongue is depressed, releasing the bolus posteriorly.

The depression of the posterior tongue follows only an instant after elevation of the apex and is the first of two such movements that occur;

the entire oral stage is ordinarily completed in a fraction of a second. It is this initial lowering of the base of the tongue that allows access of the bolus to the receptor organs in the oropharynx, resulting in the firing of the reflexive pharyngeal stage.

Pharyngeal Stage. The tongue is sufficiently posterior to be in contact with the soft palate before the latter structure moves to close the nasopharyngeal port. As the velum touches the posterior pharyngeal wall, the pharynx, which has been elevated and expanded, is now squeezed from above downward. The base of the tongue, which dropped to allow passage of the bolus, rises, and the tongue next moves backward and downward. Concurrently, the larynx is brought upward and forward, which in turn carries the anterior wall of the esophagus upward and forward through attachment to laryngeal structures. The upper portion of the tube is suddenly pulled open, with a resulting drop in air pressure. Such negative pressure speeds the bolus still more, often projecting it deep into the esophagus.

Esophageal Stage. Once in the grasp of the esophagus, the bolus is propelled by peristaltic action. The primary peristaltic wave appears to be almost a continuation of the stripping wave of the constrictors, so that the entire swallowing act, through all its stages, from original bolus formation to arrival in the stomach and reinflation of the respiratory system, is usually one continuous, synergistic, wondrously coordinated process. The peristaltic wave is preceded down the esophagus by a wave of relaxation, which facilitates progress of the bolus. This wave of relaxation also serves to relax the cardia, the orifice into the stomach, allowing passage through the cardiac valve.

Abnormal Behaviors

Tongue thrust is not a simple act, for it involves intrinsic and extrinsic lingual muscles and muscles of expression and mastication. It may occur while the tongue is at rest, during speech, or during swallowing. Since food swallowing includes all the elements of saliva and liquid swallowing, we will focus on abnormal eating behavior. Borrowing from several early writers in the field, yet not restricting our description to any one of them, we give the following "classic" description of a tongue-thrust behavior during eating.

The child takes a rather large bite of food. He chews it without making full use of facial muscles to move the food onto the grinding surfaces of the molars and onto the tongue. Instead, the tongue moves the food, first against the teeth, then later, away from the teeth. Chewing is inefficient, and no well-formed bolus results. Scattered portions of food are moved posteriorly by the creation of an anterior seal between the tongue tip and blade, anterior teeth, and one or both lips. The molars are not occluded, and the tongue remains wedged between the upper and lower teeth, all around the arch. The circumoral muscles contract, especially the mentalis; the muscles of mastication remain flaccid, and suction carries the food back to the pharynx. When the swallow is completed, the tongue has not been effectively cleared of food particles, nor have the teeth. The tongue then often carries out a cleaning procedure by pushing again against the anterior and side teeth. The basic characteristic of a tongue thrust is that the tongue contacts with some degree of force the anterior teeth when it would not ordinarily do so. Any of the other elements of the eating, drinking, or saliva handling processes may or may not be abnormal. We will describe, in greater detail, manners in which they may vary from the normal in an individual patient.

The Approach. If the tongue is low and forward at rest, as is often the case, the tongue may remain forward or even protrude to greet the approaching food. If so, it probably will return there frequently during chewing.

The Oral Stage

1. Mastication. The mores of society and the unrelenting efforts of parents have usually prevailed in achieving lip closure during mastication; however, the impulse of the deviant swallower is often otherwise. The tendency toward open lips disrupts the fine coordination required to pass the bolus back and forth over the occlusal surfaces of the teeth. The tongue tends to maul the food rather than function with the exquisite precision required in normal grinding. Instead of forming a cohesive bolus, the tongue allows dispersal of particles throughout the anterior portions of the mouth.

2. Forming the Bolus. One of the most significant characteristics may be found in the process of bolus formation. With food or saliva tucked into every cranny of the vestibule and their lingual counterparts, it is no easy matter to prepare for deglutition. Several variations are seen. Some children (and adults) display an inept and exaggerated sucking process, dropping the mandible excessively to increase the volume of the mouth and thus reduce the pressure, tensing the mentalis to maintain an oral seal, or, in some cases, contracting the circumoral muscles while squeezing excessively with the cheek muscles. Others send the apex of the tongue scampering about collecting drops or particles and retracting them into the oral cavity proper. All of this accounts for much of the facial movement that has been described, and it occurs *before* the act of swallowing.

3. Initiating the Swallow. The food having been gathered lingual to the teeth, but not necessarily on the dorsum of the tongue, the "tongue thrust" swallow occurs. Teeth may be apart or occluded, depending on the type of deviation, but tend to be apart. Lips may or may not contract excessively to resist the pressure of the tongue, depending on any number of factors such as type of deviation, status of teeth, consistency of the bolus, and how effectively it is managed. In many cases there is either discrete hyperaction of the mentalis and lower fibers of the orbicularis oris, or generalized contraction of the entire facial network.

If the teeth are separated, the tongue frequently spreads between the occlusal surfaces as far distally as the first or second molars. Since the entire tongue is displaced forward, the posterior tongue moves in an anterior and superior oblique direction until pressed with some force against the palate. Although the contact between the tongue and palate is complete throughout the length of the soft palate, the direction of force creates a stronger seal in the area slightly anterior to the junction of the hard and soft palates. This in turn forces considerable dysfunction in the oropharyngeal region and near-disuse, certainly misuse, of some of the muscles normally active.

The pharyngeal and Esophageal Stages. The bulk of the act of deglutition is encompassed in the second and third stages. Nothing abnormal has been demonstrated to occur in these phases, although Cleall (1965) observed some jerkiness and incoordination in the finer details. For the moment, and for our purposes, it is believed that we may safely ignore the pharyngeal and esophageal stages.

Post-Swallow Tongue Thrusts. During chewing and swallowing, the tongue pushes the food against and between the anterior teeth. After the swallow, a considerable amount of food often remains at the front of the mouth. The person with tongue thrust often cleans up this food with more thrusting movements of the tongue.

Summary

It is hoped that the major impression created in this section has been that of variation of function—differences in both normal and aberrant forms, with a range of behavior encompassed within each. This variability increases the importance of careful diagnostic procedures and individualized treatment planning.

History

The First Half-Century

The history of writings on orofacial myology contains more opinions than facts. Some of both will be included in this very brief, superficial summary. As is always true with new disciplines, clinical experimentation has preceded controlled research. Clinicians' imaginations conjure up exercises, technique, procedures, and approaches, which are first tried on a few patients, then applied to larger numbers, then described with numbers. If they are deemed successful, someone undertakes to apply scientific standards to the experimentation, with appropriate controls and isolation of variables.

Some of the important contributors to the development of thought and principles in the field will be mentioned in this review. For a more complete history, the reader is referred to Barrett and Hanson, 1978.

Rix

Rix published the first of a series of papers on this subject in 1946. He described a characteristic dental impairment consisting of proclinate upper incisors and a high, narrow palatal arch resulting from what he termed the "teeth apart" swallow. He described in some detail two contrasting types of normal swallowing, one consisting of the customary closure of lips, with the teeth brought into occlusal contact and the whole dorsum of the tongue coming to lie in contact with the palate. The second normal pattern occurred with soft or juicy food that may be properly swallowed with teeth apart, tongue in contact with lips and cheeks, and facial muscles contracted to form the necessary rigid cavity. He then observed that certain children utilize something similar to this latter procedure at all times, regardless of the food being swallowed, and could swallow, if teeth were closed, only with some difficulty. He noted certain similarities between some types of atypical swallowing and the suckling of infants; others he felt were "gulping" swallowers. He stated that lisps, without exception, were accompanied by abnormal swallowing.

Gwynne-Evans

A number of major contributions were made by Gwynne-Evans (1954, 1956). He was the first, for example, to suggest that something might be done to *correct* abnormal behavior. Gwynne-Evans proposed the Andresen or monobloc appliance. Rix had considered atypical swallowing as abnormal behavior. Gwynne-Evans differentiated between "abnormal behavior" and the persistence of infantile characteristics, believing that atypical swallowing was normal behavior but consistent with an earlier stage of development. He believed that muscle behavior is predetermined, patterned, and dominated by the central nervous system; that is, there is an innate plan by which groups of muscles are progressively selected, coordinated, and controlled for the service of the child's future activities. There are frequent delays in normal adjustment toward these higher levels of adjustment. Atypical swallowing represents such a delay, and some stimulus should be supplied to allow the central nervous system to resume a natural sequence of development. Since orthodontic treatment cannot influence the form of the basic bony structures, we should make every effort to assist maturation of muscle behavior and inhibit infantile characteristics, thus allowing normal growth and development to proceed.

Ballard

Ballard wrote a dozen or more articles (1948, 1951, 1959, 1960) relating to orofacial musculature. He believed that everything that occurs in a human mouth or face happens reflexively, but his use of the term "reflex" is clearly not in the usual sense. One of his preferred terms was "endogenous" to account for the posture of lips, mandible, and tongue. It is apparent that he felt adamantly that all muscle behavior is of the central nervous system and is impervious to environmental factors. Thus he rejects any possibility for educating a change in posture but asserts that many people adopt a change in habitual posture; this they do as a result of inherent physiological necessity.

Tulley

Research, primarily with electromyography and cinefluorography, occupied a large measure of Tulley's attention. Although he agreed with much of the British thinking and contributed to it, he was by far the most optimistic, concerning therapy, of those we have mentioned. Tulley

(1956) reported that his myographic studies disclosed that many adults swallow with teeth apart but that this abnormality is generally associated with malocclusion; he asserted that normal swallowing should be accompanied by firm molar contact.

Ardran and Kemp

Ardran and Kemp were two medical researchers who did a series of radiographic studies relating to deglutition. In their initial efforts they were concerned solely with normal adult deglutition. In their studies with Lind, Ardran and Kemp (1958) observed infant deglutition in both breast and bottle feeding and found no evidence that a difference existed in the mode of swallowing between the two situations. In both cases the nipple extended to the vicinity of the junction of the hard and soft palates, the tongue protruded over the mandibular gum pad, and a pumping action of tongue and mandible was apparent.

Angle

Although orthodontics had existed as a profession for some time, the first great systematizer of the field was Edward H. Angle. Most of his writing was done before the turn of the century, and the final revision of his textbook was the 1907 edition. Angle was almost more artist than dentist; he was enraptured with facial symmetry, and his language usage was delightful in portraying his passionate regard for human dentition. He described the "mischievous" tooth that nudges another out of alignment, "the majesty of pattern" of a lateral incisor, "the beautiful difference" between a cuspid and an incisor, the sacrilege when "these beautiful lines are impaired by grinding any of the marginal surfaces," which is "never anything but a travesty on Nature's beautiful normal patterns."

Angle had no conception that deglutition was a factor in malposed teeth, or indeed that the tongue was in any way responsible except as an obstruction in its resting state. He did recognize the influence of facial muscles, which he likened to hoops upon the staves of a cask. He believed that mouth breathing was the chief etiological factor, and in defining his Class II, Division I malocclusion, Angle stated that this form of malocclusion is always accompanied and, at least in its early stages, aggravated, by mouth breathing due to some form of nasal obstruction.

Angle showed models of severe malocclusions and followed with a remarkably accurate description of tongue thrust. He states:

We are just beginning to realize how common and varied are vicious habits of the lips and tongue, how powerful and persistent they are in causing and maintaining malocclusion, how difficult they are to overcome, and how hopeless is success in treatment unless they are overcome.

Rogers

Probably the first writer to propose modification of orofacial muscles was Alfred P. Rogers (1918b). He was convinced that general body posture, and particularly an imbalance of numerous groups of facial muscles, resulted in malocclusion. A personal friend, Salzmann (1957), credits Rogers with suggesting corrective exercises to develop tonicity and proper muscle function in 1906; however, his first published paper on the subject was in 1918. He made an early convert of Lischer (1912), who later dubbed the program "myofunctional therapy," a term still widely used.

Lingual habits did not occupy a major position in Rogers' thought. He never became aware of deglutition as a possible deforming influence, and suggested no program related specifically to the swallowing act. He did find many cases of open bite which he thought were the result of tongue sucking.

Rogers is perhaps best known for his concept of muscles as "living orthodontic appliances" (1918a). Although he sanctioned and practiced mechanical treatment, he strongly urged that only the lingual wire appliance be used and condemned the cumbersome expansion appliances, together with their underlying philosophy. He cited cases in which myofunctional therapy alone was responsible for correction of malocclusion and stated that retention devices should be unnecessary.

He developed a series of specific exercises for each facial muscle, to accomplish which he devised a number of ingenious bite plates, rubber exercise straps and a metal orbicularis oris exerciser. Rogers' program was widely disseminated, Ballard reporting that it was part of his early training in England.

The Truesdells

The first to interject the idea that deformity arose primarily due to forces exerted during the act of swallowing were Truesdell and Truesdell. In a paper read at a meeting of the Angle Society of Orthodontics in 1924, they proposed that some persons with severe malocclusions appeared to have great difficulty in swallowing, and likened the swallow to a gulp. In their only published article, Truesdell and Truesdell (1937) described three classes of abnormalities: (1) abnormalities immediately preceding deglutition, (2) abnormal function of the lips and jaws, and (3) abnormalities in which the tongue is involved as well as the lips and jaws.

The first consisted of "drawing the saliva," using the tongue to suck saliva from the vestibule or to scoop it up from the floor of the mouth. This resulted in a deforming pull on the muscles of expression, although the swallow itself may have been normal. This was the most common abnormality they found and was corrected merely by telling the patient to stop doing it.

The second class of abnormality, and the next most frequent, occurred when teeth were slightly parted during deglutition, throwing excess strain on the lip and cheek muscles to create the suction for swallowing. Correction was to be achieved by having the patient tense the masticatory muscles rather than facial muscles.

The third and rarest type involved a "rigid tongue," the tip of which was behind the upper incisors, the dorsum held away from the palate but in contact further back and at the sides. The chief problem in correction was to get the patient to close the teeth. Since commands proved of no avail, they devoted 5 to 15 minutes to a frontal assault. The patient was to lean forward, elbow on table and resting his chin on his closed hand, thus keeping teeth in occlusion. He was then to swallow as naturally as possible, with no effort made on his part to force the tongue to press anywhere, to think about something else and let nature take over. The Truesdells (1937) reported that as the resulting relaxation of the lips spread throughout the body, the patient began to feel drowsy. The further from normal the swallowing had been, the more complete would be the relaxation after correction.

Strang

In 1949 Strang published the opinion that every malocclusion represents a denture under the influence of, and stabilized by, balanced muscular forces that are inherent in the individual and cannot be changed by any known means of treatment.

The 50's: A Quiet Decade

During the 1950's, many orthodontists proceeded as though the problem did not exist while others set about devising the multitudinous mechanical devices to combat tongue thrust. Perhaps it is significant that after Rogers, no article appears describing these contraptions, many of which were barbaric, painful, inhuman, and ineffectual; the rationale seemed to be that the dentist was not punishing the child, only fighting some malicious entity that dwelt within the child's mouth.

For the most part, the whole idea was put aside. It was there in the background, like next year's income tax, not brought to consciousness and examined unnecessarily. Influence was certainly felt from the drastic changes in orthodontic philosophy that were thrust on an unwilling profession during this time. The original concept of orthodontic treatment was based on static position of the teeth. A bit later, "function" became the watchword, and all efforts were directed toward expanding the dental arches to achieve a functional climate in which growth of the mandible was stimulated, changes were wrought in the size and relation of the condyle, etc. Failures were attributed to uncooperative patients, particularly in wearing retention devices for several years after treatment. Roentgenographic cephalometry appeared in the 1930's, and studies conducted by this means revealed the shocking discovery that the orthodontist was not effecting the changes he had previously claimed; any osseous changes that occurred were found to be the result of natural growth processes and were unrelated to orthodontic pressures. Despair was general. According to Brodie (1950), environmental factors and oral habits were disregarded. Heredity was blamed for malocclusions; correction of the defects was not possible.

Out of the consternation and confusion came new insight, longitudinal studies of growth, and the development of the present philosophy of taking advantage of growth and development.

Brodie

While recognizing the problem, some authors toward the end of this period, even as now, probably did more to hinder than to help. Brodie (1950, 1953) maintained that growth and orthodontic treatment would solve all problems, but still managed to put in a plug for spurs on the lingual surface of incisor bands and a rubber band held between the palate and the dorsum of the tongue. We may also note Whitman (1951), who presented a somewhat oversimplified process of curing all ills by having children hold a Life-Saver® on an otherwise unidentified portion of the posterior tongue. He blithely stated that the bite would close, and if the patient had had a lisp before, it would disappear also. How many orthodontists must have kept the vigil in vain! Whitman was heard from again in 1964, but the intervening years were without noticeable effect.

Klein

A refreshing change appeared when Klein (1952) presented a well-prepared and provocative paper, first read before the Rocky Mountain Society of Orthodontists in 1951, which is still worth reading. He ranged through abnormal pressures from Chinese bound feet to the bowed legs of cowboys, discussed swallowing specifically, and in a long list of conclusions included the following:

1. Living bone is extremely susceptible to the guidance and influence of pressure and stimulus.

2. The extent to which living bone can be changed with pressure or stimulus is controversial. However, even the most conservative group will agree that alveolar bone can be changed and the teeth in that bone regulated with orthodontic treatment (planned, intentional pressure).

3. Abnormal pressure habits (unintentional pressures) also change alveolar bone and regulate teeth in that bone because the bone-building cells on the receiving end of pressure or stimulus cannot differentiate whether that pressure or stimulus is intentional or unintentional.

4. Since changes take place in living bone whether the stimulating factor is intentional or unintentional, one cannot deny abnormal pressure habits as an etiological factor in malocclusion without denying the accepted principle of planned orthodontic treatment. . . .

5. It is during the transition from the deciduous to the permanent arch that much damage takes place, and it is during this transition stage that the avoidance of all abnormal pressure habits is of the utmost importance . . .

6. The orthodontist and the patient can suffer no possible detrimental effects by eliminating abnormal pressure habits. It is logical to eliminate everything that aggravates malocclusion, everything that nullifies the plan of orthodontic treatment, and everything that is a potential factor in causing treated orthodontic cases to relapse.

7. Prevention of malocclusion is a responsibility that must be accepted by the family dentist as well as by the orthodontist.

The year 1960 may be considered as the formal opening of the present stage of thinking. Exploration was being made, activity was picking up, and thoughts were falling into place during the five years preceding this; specific pressures exerted by the musculature were being measured; radiography provided an intriguing new vista; a number of people, including Straub, Moyers, Harrington, and Barrett, were experimenting with corrective measures. The Janns visited England and returned with a full cargo of British concepts, which they successfully transplanted. The literature resulting from all of this activity began to pour forth about 1960.

The 60's

Straub

The man who must remain unchallenged as the Paul Revere of deglutition was Walter J. Straub. He raced about the countryside alerting orthodontists to the impending devastation threatened by malfunction of the tongue, commanded their attention, and forced consideration of the problem.

He first presented his theory of bottle feeding as the sole etiological factor, and a description of "perverted" swallowing in 1951. However, the time was not ripe. He suggested that existing methods be used for correction rather than offering his fellow orthodontists anything new, and his paper was largely overlooked at the time. Straub, however, was not to be denied. He employed an anonymous speech therapist, later to become a series of therapists, to work behind the scenes at his office and train his patients to swallow correctly. He compiled a great number of photographs and records for some 500 patients, made a movie on deglutition, compiled a syllabus outlining therapy procedures, and went on the lecture circuit. Starting in 1958, he appeared at dental meetings throughout the United States and even carried his message abroad. He would lecture on his theories, show his movie, together with the film of a postsurgical facial carcinoma case, and distribute copies of his syllabus to all present so that they might immediately begin to correct perverted habits. He enrolled interested dentists, speech therapists, nurses, etc., for short courses at his office near San Francisco, wherein the same materials were expounded and expanded. Part I of his trilogy on "Malfunction of the Tongue" appeared in 1960.

Straub's influence has been tremendous. He stirred a large segment of his profession from a monolithic lethargy regarding management of oral habits. It is possible that the presently intense national concern about matters deglutitory might still be a submerged fumbling without his enthusiastic presentation. Despite the growling arousal of those who would feel more comfortable were the matter still slumbering, knowledge and service can be eventually improved only as a result of such increased attention.

Straub made many original contributions to the practical literature, some valid, some misleading, some erroneous, but all with definite assurance. He was the first in modern times to incriminate conventional bottle feeding of infants.

He grouped abnormal patterns into a classification of types, and devised a much-copied therapy program.

Harrington

Up to this point every author cited has been an orthodontist. The first speech clinician to be reckoned with is Robert Harrington. If this discussion were strictly a review of the literature, we would probably omit Harrington, since he published only one article, and presented few papers. However, he was actively at work on deglutition in the Los Angeles area during the late 1950s. Since we are in private practice, we feel some empathy for his failure to find time to write. Perhaps many other speech pathologists were also engaged in similar work at the same time, but Harrington was sufficiently precocious to occupy a place on the first panel that presented the "orofacial muscle pressure imbalance pattern" to the national convention of the American Speech and Hearing Association in 1960.

Harrington staked out early the domain of deglutition as the private province of the speech pathologist, seeking dental assistance only in diagnosis and prognosis, not in therapy. He had made much of the suckling-sucking distinction. He rejected the term "habit" when applied to abnormalities. Harrington and Breinholt (1963) believed that whether abnormal oral habits were the cause of structural problems or the result of them, the habits represented the most efficient method of the organism for accomplishing the acts of chewing, swallowing, and talking.

They specified three etiological factors; (1) chronic nasal congestion, with its accompanying disruption of velopharyngeal function; (2) thumb and finger sucking, encouraging improper use of facial musculature; and (3) faulty eating habits, involving insufficient mastication of a too soft diet or flushing of coarse foods into the gullet with gulps of liquid.

The 70's and 80's

From the late 1970's to the present time, research has demonstrated that tongue thrust tends to persist into adulthood unless corrected, and that the correction of tongue thrust results in significantly less orthodontic relapse. Many orthodontists in the United States have experimented with appliances that facilitate maxillary or mandibular growth by keeping tongue, cheek, and/or lip muscles away from the teeth and their supporting structures. Anachronistic lingual spurs are periodically reinvented or reintroduced. There has been an obvious increase in the ranks of competent orofacial myologists, and dental specialists outside the traditional orthodontic fraternity are now seeking their services.

One of the more positive signs has been the continued growth of the International Association of Orofacial Myology, and the publication by this organization of the International Journal of Orofacial Myology. This organization has done much to encourage research and writing in the field.

Prevalence

Due to the undifferentiated muscle activity of lips, tongue, and jaws during sucking and swallowing, tongue thrust during infancy approaches 100% prevalence. No studies are found in the literature that would determine the prevalence between infancy and age four years. Hanson and Cohen (1973) began a longitudinal study of four-year-olds that continued until the subjects were 18 years old. Number of subjects at the beginning of the study was 225. At age eight, 178 remained in the study. At age 12 (Hanson and Hanson, 1975), 92 were available for observation, and at age 18, 61 subjects of the original 225 were seen. No treatment of any kind was recommended or discussed during any of the visits. All subjects who could be located and who would agree to be seen were studied at each of the periods of the research. Prevalences found were: Age four: 57.9%. At five years: 42.8%. Six years: 51.7%. Seven years: 35.4%. Eight years: 35.0. Twelve years: 48.9% (when subjects at this age whose tongue thrusting patterns might be termed "transitional," because of their apparent relationship with missing teeth, were eliminated from calculations, the remaining prevalence at 12 years was 38%). At age 18 years: 42.6%. (Andrianopoulos and Hanson, 1987). This longitudinal study, then, showed a decline in prevalence from age four through the age of eight, with slight rises at age six, when incisors are typically missing, and at age 12, when cuspids are often missing. Even discounting the missing cuspids, though, prevalence went from 35 percent at age eight to 38 percent at 18. Opinions widely expressed in the literature over the years that tongue thrust self-corrects after the mixed dentition period and through adolescence were not supported by the Hanson research. An examination of the data, however, showed that tongue thrust in any given subjects did not appear as a consistent pattern over the years. Only four of the 26 who were tongue thrusting at 18 years had also been diagnosed as tongue thrusters at both 4 and 8 years. A total of 12 of the 18-year-old tongue thrusters had also been tongue thrusting at age 4. Eighteen of the 61 had been tongue thrusting at the age of four, but were swallowing normally at age 18. Only 7 of the 61 examined at age 18 were classified as normal swallowers at ages 4, 12, and 18. The study demonstrated the plateau in prevalence of tongue thrust after age seven, and the inconsistency of patterns within subjects at different age intervals.

Hanson and Andrianopoulos (1987) saw 24 adult subjects (12 females, 12 males) ranging in age from 35 to 52 years to examine relationships among general dental health, periodontal health, and tongue thrust. The subjects had been randomly selected from a population of university employees. Nine of the 24 (37.5%) exhibited a tongue thrust pattern. The research was done as a pilot study, the numbers are small, but from our experience we would expect this prevalence to be upheld in a larger study. In other words, there is little difference, we expect, between prevalences of tongue thrusting at ages 7 and 50 years.

Relationships with Dental Occlusion

For over a quarter of a century, orthodontists have been referring their patients with tongue thrust for therapy, partly to facilitate the work of the orthodontist, but principally to help prevent occlusal relapse following orthodontic treatment. In 1981 Uhde examined 72 orthodontically treated patients a minimum of 12 years after treatment. He found a tendency for overjet, overbite, maxillary and mandibular arch widths, and maxillary and mandibular arch crowding to return toward their pretreatment statuses during the posttreatment period. Uhde reported finding "unacceptable" occlusions in half the subjects twelve years posttreatment. One of the factors suspected of contributing to this relapse is tongue thrust.

A review of the literature relating dental occlusion to tongue thrust is found in Andrianopoulos and Hanson (1987). Significant positive correlations have been found between the presence of tongue thrust and openbite, overjet, overbite, posterior lingual crossbite, and bimaxillary protrusion (Rix, 1946; Rogers, 1961; Werlich, 1962; Hanson and Cohen, 1973; Lamberton, 1980). Several animal studies (Negri and Croce, 1965;

Harvold et al. (1973, 1981; Miller, 1982, and Bernard, 1987), have found relationships between tongue size and posture and anterior openbites. A recent study (Andrianopoulos and Hanson, 1987) found therapy for tongue thrust to be effective in limiting orthodontic relapse. Thirty-four subjects, ages 16 to 30 years, all of whom had worn braces and an upper arch retainer, and all of whom had been out of retention for at least one year, were examined. The seventeen experimental subjects had received therapy for tongue thrust; the seventeen controls had not received therapy. All subjects were randomly selected from orthodontists' records and from therapy files at the University of Utah. All had had Class II, Division 1 malocclusions prior to orthodontic treatment. Three of the 17 therapy subjects and 12 of the nontherapy subjects were found to be currently tongue thrusting. Mean relapse in overjet since the removal of braces was 0.56 mm for the therapy group and 1.94 mm who had not received therapy. The relationship between tongue thrust therapy and the amount of relapse was found to be statistically significant at the .02 level of significance. When the therapy and nontherapy groups were combined, those subjects who were predominantly mouth breathers were found to have greater relapse (mean 3.0 mm) that those who breathed principally through the nose (mean 1.3. mm). There can be little doubt about the ability of lingual forces to create and/or maintain anterior and lateral malocclusions.

The Hanson and Andrianopoulos study of middle-age subjects (1987) investigated interrelationships among tongue thrust, malocclusion, and periodontal health in a group of 24 patients with a mean age of 43 years. They found a significant relationship between type of occlusion and presence of tongue thrust. Of nine subjects with tongue thrust, only one had a Class I occlusion. Four (20.8%) had a Class II, one (4.2%) a Class III, and three (12.5%) had mixed occlusal relationships. Among the 15 subjects without a tongue thrust, 13 (54%) had a Class I molar relationship, only one (4.2%), a Class II, and one (4.2%) a Class III. A significantly greater number of subjects with Class I were non-tongue thrusters. A significantly greater number of subjects with Class II were tongue thrusters. Collectively, a significantly greater number of persons with malocclusions, based on the Angle classification system, were tongue thrusting. (No statistically significant relationships were found between tongue thrust and periodontal or dental health, nor between malocclusion and periodontal or dental health.)

Relationships With Speech

Most children with tongue thrust have normal speech. A significantly greater number of them, though, dentalize one or more consonants that are normally produced with the front of the tongue against the alveolar ridge. These are the s, z, t, d, n, and l sounds. Only the s and z *sounds* are defective when this occurs, however, unless the anteriorization of the tongue is really extreme. When the tongue contacts a major portion of the lingual surface of any of the anterior teeth during the production of the s or z, or protrudes between the upper and lower teeth, the speech defect is called a frontal lisp.

This fronting of the tongue is probably related to the habitual forward resting posture of the tongue in people with tongue thrust. It is easier to leave the tongue against the front teeth, when it rests there, for the production of the linguoalveolar sounds, than it would be to lift it and move it posteriorly. Also, most tongue thrusters have an anterior malocclusion, such as an openbite or overjet, that creates a vertical, horizontal, or combined vertical and horizontal space between the upper and lower front teeth. In the production of the s and z, the air stream must pass through a narrow aperture at the tip of the tongue. The anterior malocclusion often makes it easier for this aperture to be created farther forward, against the teeth.

There are a number of other interrelationships among the trio of defective speech, tongue thrust and malocclusion; these will be discussed in more detail in the following chapter. Some of them provided the original basis for the entry of speech pathologists into the area of orofacial myology. Other facets attracted dentists. A few strands of their interweaving still have not been traced with perfect clarity.

Etiologies

An entire chapter in this volume is dedicated to this topic. Several causes for tongue thrust have been proposed, most of which have merit in the histories of individual patients. Since all babies apparently tongue thrust, the search for causes must be directed toward factors that keep the infantile patterns from yielding to the more mature type of swallow that the majority of people switch to by the time they reach the age of six.

Some of these proposed causes are bottle feeding with an unsuitable nipple, prolonged absence of anterior teeth, genetic tendencies toward discrepancies in palatal arch size or shape, or in intermaxillary relationships, digit-sucking habits, chronic mouthbreathing, orthodontic treatment, and oral sensory deficits.

Since the transition from tongue thrusting to normal swallowing involves developmental factors, the search for cause should uncover reasons for alterations in normal developmental patterns. Generally speaking, any factors that might restrict the amount of space available to the tongue, or that might encourage a low, forward posturing of the tongue, should be suspected of contributing to the perpetuation or development of the tongue thrust. These include enlarged tonsils, enlarged adenoids (these encourage mouth breathing), a narrow palatal arch, lingually-inclined anterior teeth, and lingual crossbite. Anterior spaces, either intra-arch, or inter-arch, between teeth, are also associated with the retention of thrusting patterns.

Treatment

Chapter Ten presents a discussion of the philosophical bases of our therapy. A variety of techniques may be available to the clinician by which to achieve a given subgoal of treatment. We believe that the choice of procedure is of far less importance than the philosophy that guides the choice, and the resulting attitude of the therapist in presenting it. Methods and exercises come and go; a good attitude, and a logical philosophical orientation—these should be steadfast.

Individualized Therapy

The goal of therapy, of whatever sort, is to modify the behavior of the patient. What these changes are, specifically and in some detail, must be ever in the therapist's mind. It is undesirable to put children into a mold, to force their conformity to a set standard; instead, we have been trained to meet the patient where he is, to assess his present abilities, and to proceed from there. After assessing a number of tongue thrusters, it became evident that their "present abilities" bore little resemblance to normal deglutition in any given particular. For a time thereafter, the components of normal swallowing were scrutinized instead. It was found that a basic group of abilities is essential to this act, that these abilities are lacking almost in toto in tongue thrusters, and that once these skills are acquired, regardless of what else the patient might or might not be capable, some hope appears of establishing normal function. The program presented here is thus oriented toward the goal, not the origin; whatever the original swallowing type, or the status of the patient, we wish to achieve normal deglutition, and we move as directly as possible toward that end. The clinician must be alert, knowledgeable, and sufficiently flexible to adapt the program to fit the specific needs and abilities of the individual patient.

Total Therapy

A tongue thrust pattern consists of many elements: Resting postures of lips and tongue; bite size; chewing, collecting, and swallowing food; postswallow cleaning-up procedures; drinking from glasses, cups, bottles, fountains, and through straws; drinking one sip at a time and drinking continuously; saliva collecting and swallowing; and tongue-fronting during speech. If treatment is to be successful it must modify all patterns that are abnormal. To fail to attend to any portion of the habit is to leave a weak link in the chain and to invite regression to former patterns as time passes.

Certainly, over the years, we have become ever more convinced of the critical influence of resting posture, the importance of having tongue, lips, mandible—every component of the orofacial complex—integrated into an easy, functional, attractive and *habitual* resting state. This forms the protective structure within which dwell the various aspects of the swallowing act itself. Failing such fortification, newly-acquired patterns of deglutition are vulnerable to a number of predators.

Efficiency of Therapy

The success ratio for therapy drops as treatment is prolonged. Accordingly, we try to disencumber the program of all nonessentials. Unless a procedure makes a direct and positive contribution to the creation or strengthening of normal patterns, it is better not assigned.

An Interprofessional Problem

The Therapists

Among specialists whom we have trained to do an effective job of therapy are orthodontists, dentists in general practice, pedodontist, speech clinicians, dental assistance, dental hygienists, and physical therapists. Each of them brought to training knowledge and experience in at least one important area of the problems to be treated. All needed to have their fields of knowledge broadened considerably before they were able to see the total child with the total problem and deal effectively with it.

Dentists. The dentist, with knowledge of oral anatomy and the developmental aspects of the problem, is in a particularly advantageous position to administer oral myotherapy. However, overhead expenses in the office usually make his/her personal participation in the therapy impractical. Most of the dentists whom we have trained have also had their assistants or hygienists trained and have merely supervised their work. In other cases, the dentists have worked directly with the patients long enough to feel comfortable with therapy and then trained their own assistants to do the work. The growing emphasis in the profession of preventive dentistry makes attention to this problem very appropriate. Throughout the history of the emergence of oral myology as a profession, the dentists have without a doubt played the key role. It is in the dentistry journals that the vast majority of clinical and research reports are to be found.

Dentists are in a uniquely favorable position to motivate the children to carry out those instructions which will help to assure their having attractive and healthy teeth. Therapy can readily be administered using a dental chair. It takes very little outlay to supplement the equipment and material normally found in a dentist's office with the special items needed to carry out therapy for tongue thrust. Many dentists, particularly in smaller communities, find themselves with no available resource person adequately trained in oral myotherapy, but with patients whose needs are critical. They have found it worthwhile, and even pleasant, to administer the therapy themselves. The general dentist, pedodontist, orthodontist, and periodontist all have vested interests in abnormal oral habits.

Speech-Language Pathologists. The speech-language pathologist is usually able to carry out a full or part-time practice with a relatively small investment. Equipment needs are minimal; therefore, daily operating overhead is small compared to that of the dentist. S/he can usually carry out the treatment using the same office and furniture that s/he uses in work with other disorders. His/her training allows treatment of speech problems that incidence studies have found to occur in 25 to 40 percent of children with tongue thrust. Although s/he requires supplemental training to be prepared to identify dental anomalies, with this training s/he is able to provide a more complete habilitative program for the patient than any other professional.

Dental Hygienists and Assistants. Dental hygienists and assistants are very adaptable to the training required to become oral myologists. Although their training and experience limits them to knowledge of the dentition, their experience in working with patients has been found to be helpful. It is convenient for the dentist to have these people see his patients because they are already working in the dentist's office. Referrals outside the office to another specialist are not necessary. The dentist is able to observe firsthand the progress of therapy and the attitude of the patient. Since most referrals for oral muscle training come from dentists, some patients who might not ordinarily act on the suggestion of the dentist to see a therapist in some other location, might do so if the person who provides the treatment is in the dentist's office.

There are disadvantages to this arrangement however. There is an air, often a characteristic odor, as well as a certain feeling of apprehension associated in the mind of the patient with a dentist's office. This disadvantage can be compensated for by having the therapist work in a separate room furnished with nondental type of furniture and designed to create an atmosphere of its own. In our experience, motivation is also a little easier when the dentist's office and myotherapy room are in separate locations. We customarily advise our patients in as positive a manner as possible that repeated failure to carry out assignments adequately leads to dismissal from therapy. It is a little more difficult to comply with these conditions when the therapist sees the patient periodically as he visits the dentist in whose office the myotherapy is conducted. Neither of these disadvantages need be serious, however, and we are acquainted with several hygienists and dental assistants who work very effectively in the dentist's office.

Orofacial Myologists. Several universities offer training courses in tongue thrust and related disorders, but few of the programs make this training mandatory for the completion of a degree. At the present time, we estimate that over 90 percent of the clinicians or therapists working with oral muscular disorders received important specific training in this area *after* receiving their professional degrees. An orofacial myologist should be an individual who:

1. Knows the anatomy and physiology of the oral region
2. Understands the vegetative and communicative functions of the teeth and orofacial musculature

3. Has an adequate understanding of developmental processes, including dental, physiological, and speech development
4. Has training and experience in the field of human motivation
5. Understands basic principles effective in changing deep-rooted habits
6. Has had appropriate training and supervised experience in treating oral myofunctional disorders

Unless, at some time in the future, a specialized degree is obtainable to prepare a person to work in this area, we believe that orofacial myologists should hold a degree or certificate in one of the specialties previously discussed. The International Association of Orofacial Myology, formed in 1972, has a qualifying examination which its applicants must pass in order to become certified members. The examination is comprehensive, and the applicant must, in order to complete the process, demonstrate clinical competence to one who already holds certification. The organization has been an effective agent in developing and maintaining standards of excellence in the profession.

The Team Members

Orthodontists

Whoever provides the therapy for the patient with an oral myofunctional disorder should do so with the cooperation of the orthodontist. Each specialist can be of immense help to the other. In our experience, a great majority of the referrals come from the orthodontist, who has already told the patient that without therapy for tongue thrust, orthodontic treatment would almost certainly fail. Most of the orthodontists who refer to us refuse to begin their treatment with a patient with tongue thrust until they receive a final report from the oral myologist stating that the patient has habituated correct muscle patterns. This requirement motivates the patient to cooperate in the myotherapy. The therapist, in turn, encourages the patient to do well in therapy in order that a satisfactory report can be sent to the orthodontist to avoid any delay in the initiation of dental treatment. The orthodontist is, through the communications received from the therapist, provided with a useful index of the degree of cooperativeness of the patient and of the parents. In addition, treatment timing depends on factors that both specialists are required to assess.

Oral Surgeons

Intermaxillary relationships in some patients are such that surgery is required to reposition the maxilla or mandible anteriorly or posteriorly. Most surgery of this type is not completed until the patient is in the upper teens. In the meantime, however, orthodontic work may be necessary to ready the patient for such surgery. Therapy for tongue thrust is often deemed essential by orthodontists before their work can begin. Conferences among concerned specialists are useful in determining the most favorable timing and sequences for the work of each of them.

The oral surgeon may be asked to examine the patient to see whether a lingual frenectomy is necessary to free the tongue for normal functioning. Infrequently, the oral surgeon determines that the tongue is too large, and performs a partial glossectomy. Both types of surgeries are relatively rare currently, however.

Other Dental Specialists

The periodontist and general practitioner both see patients who are very young. Their role in referring patients for examination and treatment by the orofacial myologist is of great importance. Particularly when anterior malocclusions begin to emerge as a result of the tongue thrust, the orofacial myologist can provide some preventive therapy, even to a child as young as four years.

Some orofacial myologists receive numbers of referrals from periodontists, who find that their efforts to provide better supportive structures for the teeth are opposed by untoward tongue and lip resting postures.

Related Oral Habits

There are numberless things that children do with their mouths that seem to us to be contributory to speech disorders and dental problems. Some contribute also to the development of psychological problems, and many seem to accompany, or be the result of, emotional disturbances. Since the specialist who works with tongue thrust needs to be trained to alter behavior that occurs at a subconscious level, involving oral musculature, it seems logical the s/he should also be ready to treat other deeply rooted habits as well. We have dealt with the following problems with some degree of success. All these problems are amenable to treatment, and the training of an orofacial myologist should include attention to each one.

Dentalization of Consonants

The linguoalveolar consonants (/t/, /d/, /n/, /l/, /s/, and /z/) are often produced linguodentally by children with a tongue-thrust problem. Particularly frequently involved are the sibilant sounds. The voiced and whispered forms of the /th/ sounds, normally produced by approximating the tip of the tongue and the posterior biting edge of the upper and lower central incisors, are often produced with an exaggerated tongue protrusion.

Thumb or Finger Sucking

A few children suck their thumbs in a way that does not seem to be detrimental to the occlusion. The majority, though, exert a relatively constant and heavy pressure against the anterior teeth when they do so. Certainly a normal swallow is not possible so long as the finger or thumb is in the mouth.

Lip Licking

This habit is self-perpetuating. the licking coats the lips with saliva, which dries, becomes sticky, and fosters repeated licking. Children who come to us for treatment very often have chapped and cracked lips that usually become healthy again when the lip licking ceases. On its way to the lips, the tongue pushes against the teeth. When the tongue is repeatedly protruded, we believe it is encouraged to remain in the relatively forward resting position.

Lip or Cheek Biting

This habit if often unilateral and results in asymmetry of the dental arch.

External Pressure Applied to the Mandible or Maxilla

Some children and adults habitually lean against their hands or fists, with the elbow on the arm of a chair or on a desk. Others push against the outside of the jaw with an object, such as a pencil or a telephone.

Tongue Sucking

We are not sure of the effects of tongue sucking on the arch configuration or on the anterior teeth. We know that it often involves the anterior positioning of the tongue against the teeth.

Habitual Biting or Resting of Objects Between the Upper and Lower Front Teeth

Some children hold a pencil, eraser, or other object between the upper and lower front teeth habitually. If the object is made of a hard material, the habit may damage the cutting edge of the teeth. If the object protrudes into the oral cavity, it may affect the resting posture of the tongue and hence affect the manner of swallowing.

Bruxism (Teeth Grinding)

When teeth grinding occurs habitually over a long period of time, it has a detrimental effect on the chewing and grinding surfaces of the teeth.

Mouth Breathing

Mouth breathing tends to be accompanied by a habitual, forward, low posturing of the tongue; research has found it to be related to the presence of tongue thrusting.

Abnormal Lingual and Labial Resting Postures

When the lower lip is habitually positioned lingual to, and pressed against, the upper incisors (as in many cases of extreme overjet), or when the tongue rests against the front teeth day and night, a light, continuous pressure is exerted against the lingual aspect of the anterior teeth that is much akin to the force utilized by the orthodontist. These resting postures are more effective movers of teeth than the more powerful, but less frequent, tongue thrust during swallows.

Evaluation

Most referrals come from orthodontists or other dentists who are very capable of spotting a tongue thrust. Seldom will a patient reach the office of an orofacial myologist who does not have a tongue thrust, if that was the stated reason for the referral. The task of the diagnostician, rather than to determine whether a thrusting pattern exists, is to determine the precise nature and extent of the problem, to identify any contributing factors that must be dealt with before or during therapy, and to gather behavioral data that will give direction to therapy. In the process of gathering those data through questions and observations, the determination of the presence or absence of a problem requiring treatment will be arrived at incidentally.

A thorough evaluation includes a case history, an examination of structures, and a study of function. The case history should include any dental, medical, social, psychological, and developmental information that may be related in any way to the tongue thrust, such as early loss and prolonged absence of anterior teeth; chronic upper respiratory infections or other conditions causing nasal blockages; any social or psychological conditions that might create insecurities manifested by adverse oral habits, such as digit-sucking; any feeding difficulties encountered, or delays in speech or language acquisition, that might hinder development of the lingual musculature. These are examples of contributory factors a case history might uncover. Any conditions that are even partially remediable, and that otherwise might thwart or slow progress in therapy, should be dealt with.

Evaluation of Structures

The examiner studies interrelationships among dental arches, teeth, tongue, and lips. Basically, the tongue must be of a size that will permit easy resting and functioning within the upper dental arch. The lips should be capable of resting together effortlessly. Any deviations from normal with respect to relationships between upper and lower teeth, or to relationships among teeth within a dental arch, should be described and measured. For example, the upper molars should overlap their corresponding molars in the lower arch slightly. The anterior maxillary arch should have a rounded configuration. If the upper molars are lingual to the lower molars, and/or the anterior maxillary arch is narrow and more pointed than rounded, tongue crowding may result. Crowding or rotation of any of the teeth may interfere with proper occlusion. When the patient places the tongue against the upper alveolar ridge and bites down slowly, the tongue should fit within the upper arch without crowding.

Evaluation of Functions

As the patient sits near the examiner while the case history is being taken, the clinician should notice whether the lips are resting together or apart. If they are apart, note whether the tongue is visible, and whether it rests against the upper or lower alveolar ridge, or against the upper or lower teeth. Eating behaviors are assessed by having the patient eat a cracker or cookie. During the entire eating process, the tongue should not be visible, in normal functioning. Does the patient chew with lips apart? When the patient signals "ready," part the lips as the swallow begins to occur. If the tongue is contacting a major portion of any of the front teeth, there is a tongue thrust on that swallow. Check it several times. Repeat the lip-parting for drinking, and for saliva swallows. Notice whether the masseter contracts during swallowing, and rate the degree of contraction of the circumoral musculature.

Engage the patient in conversation and have him/her do some reading aloud if old enough to do so. Sit at an angle from the patient in order to see whether there is lingua-dental contact on any of the lingua-alveolar sounds, t, d, n, l, s, z.

Prognosis

The diagnostician should ask two questions concerning prognosis:

1. Is the patient likely to correct the tongue thrust spontaneously without therapy?
2. Is therapy likely to be successful?

Any factors that tend to crovd the tongue in any direction mitigate against a favorable prognostic answer to either question. These include exceptionally large tonsils, anterior teeth that are lingually inclined, lingual crossbite, teeth in ectopic eruption, a narrow palatal arch, difficulty with nose breathing, or a deep overbite. Prognosis for therapy is best when both child and parents are well-motivated, and the child is mature and dependable.

Treatment

Three general approaches to treatment of tongue thrust may be considered:

1. Insert some kind of appliance into the mouth, attached to the teeth, that reminds or forces the tongue to remain away from the anterior teeth.
2. Provide therapy that trains tongue and lip muscles to function properly, and that builds new patterns into automatic behaviors.
3. Use both appliances and therapy.

The treatment of choice for most orthodontists is referral for therapy. Therapy does more than train the tongue to stay away from teeth. It reduces activities of muscles that move the tongue forward and increases, in a carefully programmed and sequenced manner, the activities of muscles that lift the tongue against the roof of the mouth. It systematically strengthens those new patterns until they occur without conscious attention. Therapy has been demonstrated to be successful clinically and through controlled research.

Summary

Tongue thrust is the most commonly occurring behavioral pattern that affects occlusion. Other undesirable habits treated by orofacial myologists include finger and thumb sucking, lip biting, sucking, and licking, and tongue sucking. Tongue thrusting is normal in infancy, reduces in prevalence to about age eight, then appears to plateau. Evidence strongly suggests that it will not go away spontaneously during adolescence.

Tongue thrust is associated with various anterior malocclusions, and with lisping. Because of its interference with orthodontic work and with the stability of orthodontically-corrected occlusion, many orthodontists routinely refer their patients with tongue thrust for therapy. The clinician should conduct a thorough diagnostic procedure and base treatment upon the findings. Therapy should be individualized. Various studies have found therapy to be effective in correcting tongue thrust and in limiting orthodontic relapse.

REFERENCES

Ardran, G. M., Kemp, F. H., and Lind, J. (1958). A cineradiographic study of bottle feeding, *Br. J. Radiol.*, 31:11.

Andrianopoulos, M. V., and Hanson, M. L. (1987). Tongue thrust and the stability of overjet correction. *The Angle Orthod.*, 57(2), 121–135.

Angle, E. H. (1907). *Malocclusion of the teeth* (Ed. 7). Philadelphia: S. S. White Manufacturing Company.

Ballard, C. F. (1948). The upper respiratory musculature and orthodontics. *Dent. Rec.*, 68, 1.

Ballard, C. F. (1948). Some bases for aetiology and diagnosis in orthodontics. *Br. Soc. Study Orthod. Trans*, 27.

Ballard, C. F. (1951). The facial musculature and anomalies of the dentoalveolar structures. *Eur. Orthod. Soc. Trans.*, 137.

Ballard, C. F. (1959). Ugly teeth. *Speech Path.*, 2, 1.

Ballard, C. F. (1960). Conclusions resumees actuelles e l'auteur relatives au comportement musculair (paper). *Societe Francaise d'Orthopedic Dento-Faciale.*

Ballard, C. F., and Bond, E. K. (1960). Clinical observations on the correlation between variations of jaw form and variations of orofacial behavior, including those for articulation. *Speech Path. Ther., 3,* 55.

Barrett, R. H., and Hanson, M. L. (1978). *Oral myofunctional disorders.* St. Louis: C. V. Mosby.

Bernard, C. L. P., and Simard-Savoie, S. (1987, April). Self-correction of anterior openbite after glossectomy in a young rhesus monkey. *The Angle Orthod., 57,* 137–143.

Best, C. H., and Taylor, N. B. (1961). *The physiological basis of medical practice* (Ed. 7). Baltimore: Williams and Wilkins.

Brodie, A. G. (1950). Appraisal of present concepts in orthodontia. *Angle Orthod., 20,* 24.

Brodie, A. G. (1953). General considerations of the diagnostic problem. *Angle Orthod., 23,* 19.

Cleall, J. G. Deglutition: a study of form and function. *Am. J. Orthod., 51,* 566–594.

Gwynne-Evans, E. (1954). The orofacial muscles: their function and behavior in relation to the teeth. *Eur. Orthod. Soc. Trans.,* 20.

Gwynne-Evans, E., and Tulley, W. J. (1956). Clinical types. *Dent. Pract., 6,* 222.

Hanson, M. L., and Andrianopoulos, M. V. (1987, March). Tongue thrust, occlusion and dental health in middle-aged subjects: a pilot study. *Intern. J. of Orofacial Myology, 12*(1), 3–9.

Hanson, T. E., and Hanson, M. L. (1975). A follow-up study of longitudinal research on malocclusions and tongue thrust. *Intern. J. Oral Myology, 1,* 21–28.

Hanson, M. L., and Cohen, M. S. (1973). Effects of form and function on swallowing and the developing dentition. *Angle Orthod., 49,* 63.

Harvold, E. P., Tomer, B. S., and Chierici, G. (1981). Primate experiments on oral respiration. *Am. J. Orthod., 79,* 4.

Harvold, E. P., Tomer, B. S., Vargervik, K., and Chierici, G. (1981). Primate experiments on oral respiration. *Am. J. Orthod., 79,* 359–372.

Harvold, E. P., Vargervik, K., and Chierici, G. (1973). Primate experiment on oral sensation and dental malocclusion. *Am. J. Orthod., 63,* 494.

Harrington, R., and Breinholt, V. (1963). The relation of oral mechanism malfunction to dental speech development. *Am. J. Physiol., 166,* 142–158.

Klein, E. T. (1952). Pressure habits, etiological factors in malocclusion. *Am. J. Orthod., 38,* 569.

Lamberton, C. M., Riechart, P. A., and Triratananimit, P. (1980). Bimaxillary protrusion as a pathologic problem in the Thai. *Am. J. Orthod., 77,* 320–329.

Lischer, B. E. (1912). *Principles and methods of orthodontics.* Philadelphia: Lea and Febiger.

Lopez-Gaviot, G. W., Little, T. R., and Joondeph, D. R. (1985, March). Anterior open-bite malocclusion: a longitudinal 10-year postretention evaluation of orthodontically treated patients. *Amer. J. of Orthod., 87*(3), 175–186.

Miller, A. J., Vargervik, K., and Chierici, G. (1982). Sequential neuromuscular changes in rhesus monkeys during the initial adaptation to oral respiration. *Am. J. Orthod., 81,* 99–107.

Negri, P. L., and Croce, G. (1965). Influence of the tongue on development of the dental arches. *Dental Abstr., 10,* 453.

Rathbone, J. D. (1955). Appraisal of speech defects in dental anomalies. *Angle Orthod., 25,* 42.

Rix, R. E. (1946). Deglutition and the teeth. *Dent. Rec., 66,* 103.

Rogers, A. P. (1918). Exercises for the development of the muscles of the face, with a view to increasing their functional activity. *Dent. Cosmos, 60,* 857.

Rogers, A. P. (1918). Muscle training and its relation to orthodontia. *Int. J. Orthod., 4,* 555.

Rogers, J. H. (1961). Swallowing patterns of a normal population sample compared to those patients from an orthodontic practice. *Am. J. Orthod., 47,* 674.

Salzmann, J. A. (1957). *Orthodontics: principles and prevention.* Philadelphia: J. B. Lippincott.

Strang, R. H. W. (1949). The fallacy of denture expansion as a treatment procedure. *Angle Orthod., 19,* 12.

Strang, R. H. W., and Thompson, W. M. (1958). *A textbook of orthodontia* (Ed. 4). Philadelphia: Lea and Febiger.

Straub, W. J. (1951). The etiology of the perverted swallowing habit. *Am. J. Orthod., 37,* 603.

Straub, W. J. (1960). Malfunction of the tongue. Part I. The abnormal swallowing habit: its cause, effects, and results in relation to orthodontic treatment and speech therapy. *Am. J. Orthod., 46,* 404.

Truesdell, B., and Truesdell, F. B. (1937). Deglutition: with special reference to normal function and the diagnosis, analysis and correction of abnormalities. *Angle Orthod., 7,* 90.

Tulley, W. J. (1956). Adverse muscle forces—their diagnostic significance. *Am. J. Orthod., 42,* 801.

Uhde, M. D. (1981). Long term stability of the static occlusion after orthodontic treatment. Unpublished thesis, University of Illinois. Reviewed by Graber, T. M. in *Am. J. Orthod., 80,* 228.

Werlich, E. P. (1962). The prevalence of variant swallowing patterns in a group of Seattle school children. Unpublished thesis, University of Washington, Seattle.

Wildman, A. J., Fletcher, S. G., and Cox, B. (1964). Patterns of deglutition. *Angle Orthod., 34,* 271–291.

Whitman, C. L. (1951). Habits can mean trouble. *Am. J. Orthod., 37,* 647.

Whitman, C. L. (1964). Correction of oral habits. *Dent. Clin. North Am.,* 541.

CHAPTER TWO

SPEECH: SOME GUIDELINES FOR THE DENTAL SPECIALIST

Dental specialists should be able to recognize abnormal speech, and particularly those defects associated with dental anomalies. In order to be able to identify defective speech, they should have some awareness of normal speech development. They should know some of the effects dental changes may have on speech patterns, and when to refer a patient for a speech/language evaluation.

Very few dental irregularities *impose* defective speech; one has only to look and listen to find many cases with structural deformities who, with compensatory movements, have acquired normal speech. Of course, to make those compensatory movements, the speaker has to have good motivation, some degree of fine muscle coordination, and reasonable intelligence.

Articulation and Language

Speech is literally "hot air." It is a column of breath supported by the lungs, acted on first by the vocal folds of the larynx, and thereafter molded and reinforced by the various tissues, organs, and cavities of the throat, nose, and mouth. It is influenced and often colored by the personality and emotions of the speaker. It is emitted as a series of hums, hisses, and explosions, each produced in a distinctive way or in one of a limited number of distinctive ways. Through common agreement, stereo-typed over the years, we accept these noises as meaningful. There is nothing inherently binding in those sounds; our ancestors might have chosen other noises, and in some cases they did, in other languages or in English sounds that have been lost in evolutionary processes. The sounds we accept as normal are those which are most closely adapted to normal human structures and functions, those which can be produced most *easily* by the average adult.

39

Articulation is the production of speech sounds. These sounds are joined together into words. The words are spoken in meaningful combinations and sequences called "language."

The child learns to speak the language, and, later to write it and read it. These skills are a challenge to many children, because English is not a phonetic language. Some letters may represent several sounds or no sounds at all. Thus c may mean either k or s, which are unrelated sounds, whereas x is either ks or gz, making both c and x superfluous and ambiguous symbols. Certain combinations of symbols are absolutely inexplicable in the English language, and it has been necessary to devise a special system, the International Phonetic Alphabet, in order to carry on the work of speech correction. We won't impose phonetic symbols on you in this chapter; we will use the conventional alphabet letters.

Speech Development

Growth and development are important concepts for the speech clinician as well as for the dentist. Speech sounds are not acquired in toto, but develop in a fairly predictable fashion. Children tend to learn first those sounds that require the fewest muscles, the largest muscles, and the least degree of coordination between muscles. Based on the ability to execute gross sounds, progress is made in a somewhat physiological manner toward the more sophisticated sounds.

Very rarely does a speech defect involve vowel production; vowels are usually mastered in the early vocal play of the infant, partially because they necessitate no major interruption of the voice stream, as is required by the consonants. If vowel distortion is perceived, it is usually wise to consult a speech/language pathologist.

Arlt and Goodban (1976) studied consonant acquisition in 240 children, ages three to six years. The following norms indicate the age at which 75% of the children were consistently producing correctly each of the consonants tested:

3.0 years: m, n, h, p, ng, k, d, w, b, t, g, wh.

3 years, 6 months: v.

4 years: s, l, ch, zh, j, z.

4 years, 6 months: sh.

5 years: r, voiced and whispered th.

In their study, less than 75% of the children were producing the y sound correctly at six years. Other studies place its acquisition at four years.

Most children have normal speech by the age of five years. If a child is six, produces only one sound incorrectly, and only slightly so, there is usually no need for a referral. By the age of seven, any abnormal sound productions are grounds for treatment.

Speech Disorders

Speech defects cover a broad range of conditions, for most of which there are no dental implications—only a need for recognition. The lay person often classifies all speech defects into one category: *tongue-tie.* Diagnosis is quick and easy since it does not require oral examination. The concept is broad enough to encompass brain injury, physical defects, emotional disturbance, and faulty learning. Correction is simple: clip the lingual frenum. Far more cases have been referred to the writers for speech treatment with the lay diagnosis of "tongue-tie" than all other disorders combined. In point of fact, of several thousand such cases, precisely six have in reality suffered ankyloglossia. Several abnormal frena have been seen accompanied by normal speech, and many speech defects have been seen following frenectomy.

Identification of Speech Disorders

A quick screening test may be administered by nonspeech-language pathologists by having the child count to twenty, then engaging the child in a brief conversation. As you listen to the counting you will soon become aware that it is very common for normal-speaking children to dentalize the l in 11. Don't label that sound as defective when this occurs. The th in "three" and "thirteen" is produced by normal speakers with the tongue slightly protruded between upper and lower teeth. Unless the protrusion is greatly exaggerated, call it normal. Interestingly, the whispered th, as in "think," "throw," "both," and "bathtub" is next-to-last in frequency of occurrence of consonants in English (Miller, 1951). The voiced th, as in "the," "other," and "breathe" is the seventh most frequently occurring sound, following n, t, r, s, d, and l, in that order. As you watch a child talk who uses many articles and demonstrative adjectives, you may erroneously perceive a tongue thrust. Even if the tongue protrusion on the th sound seems excessive, watch carefully as the child produces the t, d, n, l, s, and z before making a diagnosis of tongue fronting during speech. Importantly, even habitual tongue fronting during speech, if unaccompanied by forward resting tongue postures and vegetative-related tongue thrusting, is a speech problem, not a tongue thrust.

Prevalence

You should expect that approximately 1 in every 10 children you see will have some type of speech disorder. About half this group, or 1 in 20, will display a defect severe enough to warrant therapy. These defects include poor articulation, stuttering, voice disorders, delayed language, hearing-related disorders, brain injury-related disorders, and cleft palate speech.

Causes

At least 75% of all speech defects have their origin in some sort of breakdown in the learning process. Given an intact central nervous system, normal hearing, and normal structures, faulty learning still occurs. We hypothesize that the following is a typical process for learning defective articulation. Children are stimulated by sounds they hear people make. They try to imitate them. Their attempts may not be perfect, but are understood, hence reinforced by their listeners. They accept the less-than-perfect sounds they have produced because they are understood and accepted.

In some cases of poor articulation, adverse conditions for speech learning may be operative at home. When early speech efforts are met with indifference or impatience, we should expect some delay in future acquisitions. Some parents inadvertently make most of their communication with their child of a corrective, or punitive, or negative nature. At the other extreme, parents may try too hard to speed up the child's speech development. Demands on speech may be of the wrong type, or ill-timed, or made with the wrong attitude.

A principal tutor is the speech children hear around them, patterns set by parents, siblings, and friends. Children not adequately exposed to normal adult speech are robbed of necessary models. If grunts and pointing bring wanted objects or foods to the child, motivation to learn to talk is dampened. Prolonged illnesses during critical speech-learning periods, exposure to two languages spoken in the home, or emotional traumas experienced may result in speech difficulties in some children. In other cases, children may not be capable of producing a sound correctly, try as they may. Hence they substitute the closest sound they are able to produce. The substituted or distorted sound is reinforced, and becomes habitual. After the pattern for the sound is established, it is turned over to the unconscious system as a unit, and the child is no longer aware of any difference between his/her speech and the correct pattern. S/he hears the abnormal pattern as normal.

Organic Factors

Lowered intelligence, brain injury, emotional shocks or conflicts, physical trauma, poor coordination, and dental abnormalities: all serve as possible causes of speech defects. Any of the organic conditions may exist to a minor degree in company with dental malocclusions. They may result in a dysarthria or an apraxia.

Dysarthria. When articulatory contacts are weak or incomplete, resulting in slurred speech, and the defect is due to muscle or nerve weakness or pathology, the speech is called dysarthric.

Apraxia. Patients with impaired control over motor movements due to central nervous system pathologies may have an apraxia, the inability to make a movement voluntarily that can otherwise be produced involuntarily. If the apraxia affects the muscles of articulation, it is called a verbal apraxia. The apraxic typically produces some sound other than the one intended, or some unintended word.

Common Articulatory Defects

Articulation problems constitute about three fourths of the case loads of most speech-language pathologists. An estimated 80% of those articulatory problems deal with the incorrect production of the s, l, or r sounds. In one way or another, the most common defects represent hypoactivity of tongue muscles other than the chief tongue protruder, the genioglossus.

r and l

Children who produce these sounds incorrectly fail to raise the anterior tongue adequately. Either sound may be made with rounded lips, as a w. The r is often produced as a neutral vowel, or as a very dull-sounding r, because of the lower position of the tongue. Parents often call the child, or the child's speech, "lazy." Orofacial myologists firmly believe that much of the problem is often the low, anterior resting posture of the tongue. When the child speaks, the tongue moves from its resting position only as far as it needs to to produce speech that will be understood. The establishment of a normal, tongue-up resting posture, along with the correction of swallowing patterns that bring the tongue forward rather than up, has been demonstrated to facilitate correction of these sounds.

s Defects (lisps)

Lisping comes in assorted styles and sizes, and should be of particular interest to the dental specialist. The most common of all speech defects is the frontal lisp, which is an interdental or dentalized s, sounding much like a th sound. The z sound is produced exactly like the s, except voice is added. When the s is defective, the z is almost always affected as well. In more severe cases, the sh, zh, ch, and j are also anteriorized. Tongue pressures against the teeth, or between upper and lower teeth, caused by the th, or by the anteriorized lingua-alveolar sounds, are so light that they probably do not cause teeth movement. As a part of the whole behavior pattern of anterior tongue movement, though, they are important.

Another type of s defect is the lateral lisp, which has a distinctive and very juicy sound. In normal s production, the sides of the tongue lift against the alveolar ridge and prevent lateral escape of air. The lateral lisper allows air to escape over one or both sides of the tongue. A lateral lisp can be produced by sustaining a l sound, then continuing it after turning the voice off. In other words, the lateral lisp is a whispered l sound. Many children with lateral lisps also thrust one or both sides of the tongue during swallowing. Exercises that train proper tongue resting postures, and those that train elevation of the sides of the tongue, all a routine part of therapy for tongue thrust, help considerably in the remediation of lateral lisps.

There are other types such as nasal lisps, occluded lisps, and strident lisps. These are, in order named, the emission of the s through the nostrils instead of orally, the blockage of the air stream by a tongue held too tightly against the alveolar ridge, or by overly-approximated lips or teeth, and an s resembling a whistle. Children with repaired clefts of the palate often display a nasal lisp. People with sensorineural, high-frequency hearing losses may have occluded lisps. Strident lisps result from too small an aperture between the tongue tip and the alveolar ridge, usually caused by too much tension in muscles that elevate the anterior tongue.

Dentalization of t, d, and l

When the tongue contacts the anterior teeth during the production of these sounds, only the careful listener detects any abnormal sound. They are normally produced with the tongue tip against the alveolar ridge.

Speech-language pathologists without training in oral myofunctional disorders would ordinarily not treat them. When they occur along with other patterns characteristic of tongue thrusting, they should be given attention. Children who dentalize these consonants seem to call the anterior teeth "home" for the tongue; the tongue goes home whenever it gets a chance, even during pauses in conversations. The pattern is sometimes severe enough to create a visually unattractive speech, thereby calling much attention to itself.

Delayed Speech

When vocabulary, sound production, or language are developmentally delayed enough to call attention to themselves, the child is in need of speech-language treatment. Well-meaning dental or medical specialists may mistakenly assuage the concerned parent with the response, "Don't worry, he'll grow out of it." Some of the children never grow out of it. Others do, but during the "growing out" years suffer embarrassment from the attention given their speech by their peers. The time to deal with this problem is the earliest possible moment after the child has passed the upper normal limit for any given sound as specified above. The dentist who is able to recognize the symptomatology of delayed speech, and transmit this knowledge to parents, can perform a great and valuable service to his patient.

Relationships Between Dentition and Speech

There are many situations in which the dentist is capable of rendering real assistance to the speech pathologist. Although some patients seem to have the resourcefulness and the ability to overcome dental abnormalities that affect their speech, many others do not. We will discuss in this section some conditions wherein the dentist, by modifying structure, can provide the patient with better equipment with which to produce normal speech.

A considerable amount of material has been written on the subject of dental impairment caused by faulty speech, but very little of a valid and specific nature has been said in print on the potential of dental treatment to improve speech. An excellent treatment of this topic is found in Bloomer's (1971) chapter on speech defects associated with dental malocclusions and related abnormalities.

Snow (1961) found a positive correlation between the misarticulation of certain consonants and the condition of upper central incisors. Fairbanks and Lintner (1951) made a series of dental comparisons between two groups, one with superior speech and one with inferior articulation. They found no significant difference in the height or width of the dental arch and palate, in molar or anterior occlusion, nor in degree of overjet; only open bite conditions were significantly more frequent in the inferior group.

Rathbone (1955) called attention to the need for dental recognition of defective speech, devised a test by which the dentist might evaluate speech competency, but advised referral of the patient to a speech correctionist for training.

Rathbone and Snidecor (1959) reported a study of ten patients with marked speech defects. Orthodontic treatment was instituted with not only traditional goals in mind but also some thought to effective structures for speech. Although no formal speech therapy was performed, their patients showed a reduction of faulty sounds from a mean of 6.4 to a mean of 1.5. Unfortunately, no ages were given for their subjects, so that no estimate can be formed as to the effect of growth and development, if any, or of other factors that might have influenced the result.

What changes in the patient's speech should the dentist expect to find after correction of dental abnormalities? More to the point, what dental irregularities should be assailed with expectation of improving speech habits? Note first that if there is malocclusion accompanied by speech defect, there is a strong possibility of tongue thrust.

Anterior open bite is perhaps the most rewarding condition; when uncomplicated by defective deglutition, any sibilant sound (s, z, sh, zh, ch, j) is sometimes corrected simply by closing the bite. If other sounds are defective, it would be logical to expect their improvement.

Posterior open bite or crossbite would have little effect on speech unless a lateral lisp is in evidence. Closed bite and dental spacing have little direct effect on speech nor do high or narrow palates or malposed individual teeth; the latter may interfere with connected discourse but is not associated with a specific speech disorder.

Maxillary protrusion can affect speech in several adverse ways. If incisal labioversion is severe, many patients find it anatomically impossible to occlude the lips for the bilabial plosives p, b, and m; instead these sounds are produced by approximating the lower lip and upper teeth, thus relieving the upper lip of one of its major functions and leading to its further retraction and incompetence. The resultant sound is often acoustically normal but cosmetically unpleasant. Reducing the protrusion and then instituting a program of lip exercises to restore function to the upper lip usually eliminates the speech defect.

If maxillary incisal protrusion is accompanied by gross mandibular retrusion, generalized distortion of many consonants may be noted. In these cases correction of structure should certainly have precedence.

Maxillary protrusion in Class II may also be a major factor in frontal lisping. Reducing the protrusion in this case gives no assurance of speech improvement, but some children find it difficult to perform the necessary tongue positioning to achieve sharp sibilants when the discrepancy between arches in great.

Frequently speech defects that are found in conjunction with Class III cases may be a product of malocclusion. Although some pseudo Class III cases are the product of tongue thrust, the genuine article can cause great speech disturbance. If there is contraction of the maxillary arch, a common circumstance, the apex of the tongue perforce contacts the maxillary teeth rather than the alveolar and palatal sources, thus distorting many sounds. The f and v sounds are often "inverted," that is, produced by touching upper lip to lower teeth rather than by the customary contact of lower lip to upper incisal edges. These sounds in some cases are produced by approximating the lips, which eliminates dental contact and results in obvious distortion of speech. Dental or surgical correction of the structure should be the first step, followed by speech therapy if still necessary.

The above is a fairly complete inventory of the conditions wherein the dentist may expect to render speech correction through modification of structure. Additional sounds may be improved by combining dental treatment with correction of abnormal swallowing; however, the dentist who implies to his patient that treatment will have more far-reaching effects than specified above has stepped beyond the limit of accepted practice.

A final and converse thought in this connection is that even though the denture, prosthesis, or appliance placed in the patient's mouth may not improve speech, at least it should be so designed as not to impede speech.

People without any teeth at all sometimes speak essentially normally. The flexibility of the tongue, lips, and other articulators permits extensive compensatory action.

Conclusion

A summary of normal speech development has been presented, along with a brief classification and description of abnormalities, and a discussion of the frequency of their occurrence. It was noted that in most of these conditions formal speech therapy is recommended for proper management.

Also, a resume was given of those conditions in which speech might be corrected solely through dental treatment, and a few conditions were noted for which dental treatment would not be indicated as a substitute for speech therapy.

Most speech clinicians would be delighted to give advice on the speech problems of your patients. Unfortunately, very few have had formal training in therapy for tongue thrust. This will change as momentum builds and new knowledge is gained.

It may be advisable to stress again what defective speech is *not*. It is not abnormal to see the tongue during connected speech. In particular, it is normal for the apex of the tongue to touch or extend beyond the incisal edges in forming *th* sounds. It is not the demonstration of a pernicious lip habit each time a child forms an f or v sound; the lip must contact the upper teeth, even though an effort must be made if malocclusion is severe. The muscular force exerted during normal function on the dentition is far below the level required for deformity.

REFERENCES

Arlt, P. B., and Goodban, M. T. (1976). A comparative study of articulation acquisition as based on a study of 240 normals, ages three to six. *Lang. Speech Hear. Serv. in Schools, 7,* 173–180.

Bloomer, H. H. (1971). Speech defects associated with dental abnormalities and malocclusions. In L. E. Travis (Ed.), *Handbook of speech pathology and audiology.* New York: Appleton-Century-Crofts.

Fairbanks, G., and Lintner, M. V. H. (1951). A study of minor organic deviations in "functional" disorders of articulation. 4. The teeth and the hard palate. *J. Speech Hear. Disords., 16,* 273.

Miller, G. (1951). *Language and communication.* New York: McGraw-Hill.

Rathbone, J. S. (1955). Appraisal of speech defects in dental anomalies. *Angle Orthod., 25,* 42.

Rathbone, J. D., and Snidecor, J. C. (1959). Appraisal of speech defects in dental anomalies with reference to speech improvement. *Angle Orthod., 29,* 54.

Snow, K. (1961). Articulation proficiency in relation to certain dental abnormalities. *J. Speech Hear. Disords., 26,* 209.

CHAPTER THREE

DENTAL BASICS: CONSIDERATIONS FOR THE OROFACIAL MYOLOGIST

Orofacial myology has attracted clinicians from a number of fields, equipped with a variety of backgrounds. All must work in close association with dentists; myofunctional disorders, as a rule, manifest themselves in dental problems. The therapist must be capable of understanding and discussing these dental aspects. S/he should be able to discuss them with the dentist using dental terminology.

For example, the speech pathologist-turned-myologist, accustomed to working in the same area of the body as the dentist, may feel that his/her knowledge of the involved anatomy is excellent—and from a neuromuscular point of view it may be more than adequate—but is a different knowledge. It is based on different ideas about reaching different goals and couched in different words, extending only superficially into the dental province. And so, not infrequently, the clinician and the dentist find problems in communication arising from deficits in the clinician's background. This chapter will attempt to bridge those gaps.

As an added attraction, we will include brief sketches of a few dental procedures. These are some of the more common maneuvers, each with relevance to orofacial myology, but to which the therapist may not have been formally introduced.

The Tooth and Its Supporting Structures

Dental Anatomy

The tooth is composed of four basic tissues: enamel, dentin, pulp, and cementum (Fig. 3-1).

The *enamel* is the hardest substance in the human body. It is the white, outer surface of the anatomical crown.

The *dentin* forms the body of the tooth, both crown and root. It is yellowish and is composed of a dense, calcified tissue. Much softer than

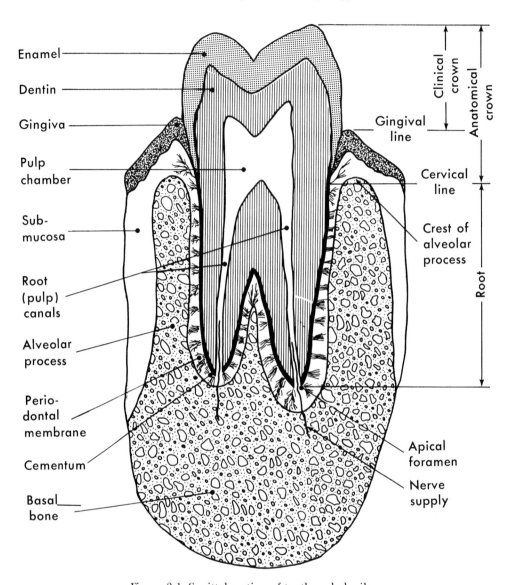

Figure 3-1. Sagittal section of tooth and alveilus.

enamel, it is not normally exposed at any point in the natural tooth.

The *pulp* occupies the pulp cavity and the root canals. It is a delicate, fibrous tissue, richly supplied with blood vessels and sensory nerves. It supplies nutritional elements to the main body of the tooth but is primarily concerned with the formation of dentin.

The *cementum* is a layer of modified bone that covers the root of the tooth. Very thin at the cervical line, it thickens toward the apex of the root.

The *apical foramen* is the tiny opening of the root canal at the apex of the root through which pass the blood vessels and nerves supplying the pulp.

Supporting Structures

The tooth is supported by four additional tissues, collectively called the periodontium: the alveolar process, the periodontal membrane, the mucous membrane and submucosa, and the gingiva (Fig. 3-1).

The *alveolar process* is a band of bone consisting of the portion of both maxilla and mandible that forms and supports the alveoli, or sockets, of the teeth. The alveolar process lies along the occlusal border of the main body of the bone proper and is continuous with it, but it is well to note that whereas the orthodontist is readily able to modify the alveolar process, the bone proper, or *basal bone,* is much less amenable to orthodontic manipulation. Basal bone is the skeletal bone that supports the alveolar bone.

The *periodontal membrane* surrounds the cementum and is composed of bundles of fibers intermixed with connective tissue. The fibers attach to the cementum on one side and to the alveolus at the other, thus attaching the tooth to its socket. The connective tissue contains not only blood vessels and lymphatics but also receptor nerves of unusual responsiveness. Differences of only fractions of a millimeter in the thickness of paper can be detected by the proprioceptors located in the periodontal membrane and in the muscles of mastication. The periodontal membrane is also viewed by many dentists today as a ligament, and the tooth-alveolus relationship is seen as a dento-alveolar *joint,* placed in motion by forces of varying pressures acting on the tooth. The basic purpose of the periodontal ligament is not only to maintain the teeth in their normal position, but also to act as a shock absorber, cushioning occlusal forces as they are transmitted to the bone.

The *oral mucous membrane* consists of two layers: the outer surface epithelium and the *lamina propria.* Between the mucous membrane and the periosteum of the bone lies the *submucosa,* a layer of connective tissue of varying thickness containing blood vessels and nerves.

The *gingiva* (gum) immediately surrounds the tooth and is a keratinized layer of tissue in which no blood vessels are visible, in contrast to the mucous membrane.

Dental Orientation

Surfaces of the Tooth

Descriptive terminology in dentistry has, of necessity, been altered somewhat from traditional anatomical terms. Because of the curve of the dental arch, a "lateral" surface (the surface farthest from the midsagittal plane) would not indicate corresponding parts of an incisor and a molar; it would describe the surface of an incisor facing a neighboring tooth, whereas it would mean the buccal surface of a molar. Special usage has therefore developed.

The point of orientation is the sagittal midline of the dental arch. The surface of any tooth that lies nearest the medial line, following the curve of the dental arch, is called the *mesial* surface, whereas the surface farthest from the median line is called the *distal* surface. Two surfaces of each tooth are thus described.

However, anterior teeth have four surfaces, and posterior teeth have five. The surface lying nearest the lips is known as the *labial* surface in the case of incisors and cuspids, while the corresponding surface of bicuspids and molars is called the *buccal* surface. Opposite this surface, all teeth have a *lingual* surface; it is becoming increasingly popular to simplify this terminology by referring to both the labial and buccal planes as the *facial* surface, while still others prefer the term *vestibular* surface. Thus the anterior teeth have mesial, distal, labial (facial), and lingual surfaces. Bicuspids and molars have additionally an *occlusal* surface: that surface which contacts the corresponding tooth in the opposite dental arch. In the case of anterior teeth this part of the tooth is the *incisal edge* rather than a surface.

The *proximal* surface is an indefinite term simply indicating the surface facing or approximating a neighboring tooth, whether mesial or distal.

Similar terminology is used to locate any line, point, or other feature of a tooth, for example, the *lingual* groove or the *mesial* pit. Where two surfaces approach each other, the area is indicated by dropping off the final letters of the name of one surface, replacing them with "o," and then combining the two names. Thus the *mesiobuccal* ridge, the *distolingual* cusp, etc., differentiate precise features.

The occlusal surfaces of posterior teeth are extremely complex; dentists of an earlier day frequently used the analogy of a "gristmill" in referring to this region. A smooth, flat surface would make trituration of many foods impossible, but simple edges would prove disadvantageous for a number of reasons. Instead the surface is so arranged as to allow the complex rotary motion of mastication with maximum tooth contact in all ranges of occlusal articulation.

Topographical Features

An illustration of the topography of the region is given in Figure 3-2. The crown of the tooth is characterized by the following principal features: pits, grooves, fossae, cusps, ridges, and inclined planes.

Pits are sharp depressions usually found at the junction of two or more grooves.

Grooves, also called sulci, are long, narrow depressions in the enamel

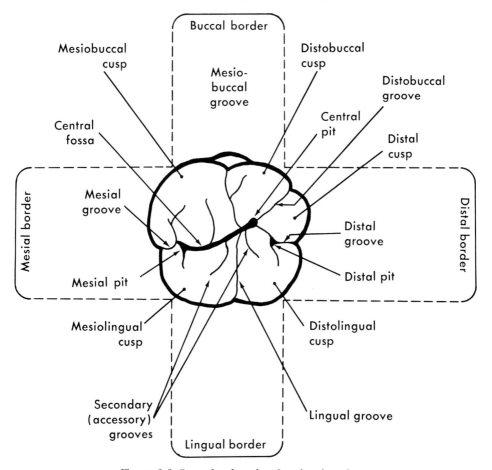

Figure 3-2. Some landmarks of occlusal surface.

that may extend over the edge of the occlusal surface onto the sides of the tooth. Some are called *primary* or developmental grooves because they are constant and are a developmental part of the particular surface on which they occur. *Secondary,* or accessory, grooves include the other grooves that occur on the occlusal surface; they are less deep and less constant.

Fossae are large depressed areas, some smooth and shallow, others deep and angular. An example of the former occurs on the lingual surface of the maxillary incisors, whereas the latter are found primarily on the occlusal surfaces, where they are formed by pits and grooves, bounded by ridges, and named for the pit they encompass. Thus the *central fossa* surrounds the central pit.

Cusps are the high points, or marked elevations, of enamel, which lead to the names of certain teeth; that is, a cuspid is a tooth with one cusp, and a bicuspid has two cusps. Molars, of course, commonly have four, but some have three or five cusps.

Ridges are long convex elevations of enamel found on all surfaces except the mesial and distal. The incisors have ridges running gingivally from the incisal edge; for other teeth, almost all ridges extend down from the point of a cusp.

Inclined planes are the sloping areas on the occlusal surface that lie between the crests of the ridges and the primary grooves. They are of various shapes and are usually crossed by secondary grooves.

Figure 3-3 shows the tooth in lateral aspect. It may be noted that the area where two proximal surfaces meet is called the *contact* or *contact point.* Because of the contour of the teeth, triangular-shaped spaces, called *embrasures,* are created between the crowns; they extend in four directions from the contact area. The occlusal embrasure is noted in Figure 3-3, as is the *interproximal space.* There are also lingual and buccal embrasures.

Relationship of Tooth to Arch

It is often necessary to describe the position of a tooth in relation to the dental arch in which it occurs, and thereby to its adjoining teeth. The system most frequently used is Lischer's. He used the proper term to indicate the direction from the normal, added the suffix "-version," and arrived at the following:

Mesioversion—mesial to the normal position.

Distoversion—distal to the normal position.

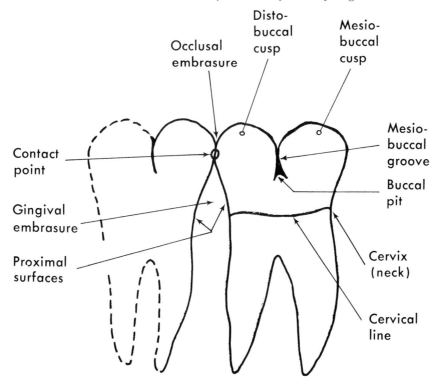

Figure 3-3. Tooth-to-tooth relationship.

Linguoversion—toward the tongue.
Labioversion (or buccoversion)—toward the lip (or cheek).
Infraversion—away from (not reaching the line of occlusion.
Supraversion—extending past the line of occlusion.
Axioversion—wrong axial inclination, tipped.
Torsioversion—rotated on its long axis.
Transversion—wrong order in the arch, transposed.*

The term "supereruption" or "overeruption" is usually acceptable in lieu of supraversion. *Intraversion* and *extraversion* indicate teeth or other maxillary structures that are too near or too far from the median plane. When a tooth is bodily malposed, it is said to be *displaced,* but if only tipped or tilted, it is mesially, distally, lingually, labially, or bucally *inclined.* If the entire incisor segment is inclined labially or lingually, it is usually called *protrusion* or *retrusion,* respectively, and when both upper and lower incisors protrude, the proper term is *bimaxillary protrusion.*

---

*From Lischer, B. E. (1912). Principles and methods of orthodontics. Philadelphia, PA: Lea and Febiger.

Dental Development

Deciduous and Permanent Teeth

The subjects of oral histology and embryology fill many volumes, at least a few of which should be surveyed by anyone purporting to do serious work with deglutition. Any treatment here would necessarily be so incomplete as to frustrate the research person and bewilder the neophyte. With the firm realization that teeth are not surgically grafted into the mouth, but are subject to the same histological and embryological processes as other tissues, we will outline the more obvious developments.

The enamel organ of the first tooth bud forms at about the fourth or fifth month of intrauterine life. These buds gradually differentiate from their genetic tissues, and calcification of the crown begins. Once the crown is complete, root starts to form, and with root comes eruption. That is, the newborn is found to have skeletal bone of the jaws proper, but only rudimentary alveolar bone, which develops only as required to encase the growing teeth. We might pause to note that the fetal tongue protrudes between the gum pads, completely covers the skeletal bone of the lower jaw, and approaches closely, or actually contacts, the lingual surface of the lower lips. This lingual posture is continued by the neonate in nursing, the tongue and lip working in mutual opposition once function begins. The environment into which the tooth is born is thus as stressful as the cruel world into which the baby him/herself is deposited. Before a tooth has erupted, the developing alveolar bone is subjected to definite shaping forces from the counteracting tongue and lip, so that the teeth must then intrude between two already veteran antagonists, the results of which form the fountainhead of this volume.

The teeth that usually erupt first are the mandibular central incisors, usually at about 7 months of age. Several authorities have noted that the time of arrival of the various teeth, providing it is within broad normal limits, is of relatively small importance; the significant factor is the *sequence* in which the teeth erupt. The usual order for the primary dentition is central incisor, lateral incisor, first molar, cuspid, and second molar, with the mandibular tooth erupting one to four months before the antagonist tooth in the maxillary arch. In most cases the primary dentition is completed between 2½ and 3 years of age.

As indicated in Figure 3-4, the deciduous teeth in each arch total 10, or 20 teeth in the two arches. The mandibular arch is normally the con-

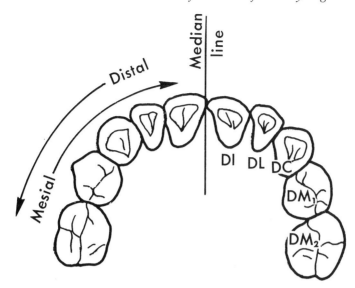

Figure 3-4. Arrangement of deciduous teeth in dental arch. *DI,* Deciduous central incisor; *DL,* deciduous lateral incisor; *DC,* deciduous cuspid; *DM,* first deciduous molar; *DM₂,* second deciduous molar. The term *distal* refers to surface, point, or movement away from median line; *mesial* implies surface, point, or movement toward median line.

tained arch, that is, the lower teeth fit within the circumference of the maxillary teeth when brought into habitual occlusion.

To simplify the identification of the permanent teeth, it is customary to start with the central incisor at the median line and number the teeth posteriorly around half of the dental arch, as seen in Figure 3-5. By then adding the prefatory terms of maxillary or mandibular, right or left, all 32 teeth are named.

Eruption

Many orthodontist feel that the most important phase in the life cycle of a tooth is the process of eruption. Little can be done about this aspect regarding the primary dentition, but elaborate natal plans often precede the permanent tooth, as we shall see in the discussion of "serial extraction." Although the process of eruption continues at a reduced rate throughout the life of the tooth, it is considered primarily as the complex of activities that carry the tooth from its developmental crypt into the mouth and into occlusion with an antagonist tooth. As the roots of a permanent tooth elongate, driving the crown through the gingiva, additional alveolar bone is deposited, and the roots of the deciduous predecessors are resorbed. Although these three processes are usually synchronized,

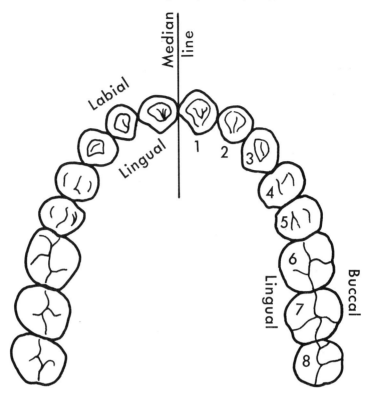

Figure 3-5. Identification of permanent teeth and their relationship within dental arch. 1, Central incisor; 2, lateral incisor; 3, cuspid (canine); 4, first bicuspid (premolar); 5, second bicuspid (premolar); 6, first molar; 7, second molar; 8, third molar (wisdom tooth). The term *labial* indicates surfaces around front of arch extending to distal surface of cuspids, whereas *buccal* describes outside area on each side that lies posterior to cuspids. *Facial* may imply either.

they may be less dependent on one another than has been thought.

As previously noted, the expected time of arrival of the teeth may vary rather widely, and many schedules have been proposed in many books. It is known that eruption is under endocrine control but may be influences by heredity, pathological conditions in the mouth, or systemic disease. The times shown in Table 3-1 are intended to show some general averages and certainly do not reflect upper and lower normal limits.

The sequence of eruption of the permanent teeth is shown in Table 3-1 and is in part a function of the exfoliation of the deciduous teeth. The normal order, under the ideal conditions that customarily prevail, is shown in Table 3-2. Many studies have shown that any other sequence leads to malocclusion; in some cases the malocclusion can thus be identified prior to its actual development.

Table 3-1. Time of eruption of the teeth

Primary dentition		Permanent dentition	
Central incisors	5– 8 mo.	First molars	5– 7 yr.
Lateral incisors	8–10 mo.	Central incisors	6– 8 yr.
First molars	10–16 mo.	Lateral incisors	7– 9 yr.
Cuspids	16–20 mo.	First bicuspids	8–10 yr.
Second molars	20–30 mo.	Mandibular cuspids	9–11 yr.
		Second bicuspids	10–12 yr.
		Maxillary cuspids	11–13 yr.
		Second molars	12–14 yr.
		Third molars	17 yr. up

Table 3-2. Normal sequence of eruption

Mandible	Maxilla
1. First molar	
2. Central incisor	
	3. First molar
4. Lateral incisor	
	5. Central incisor
	6. Lateral incisor
7. Cuspid	
8. First bicuspid	
	9. First bicuspid
	10. Second bicuspid
11. Second bicuspid	
	12. Cuspid
13. Second molar	
	14. Second molar

Mixed Dentition

The mixed dentition stage is the period of time during which both deciduous and permanent teeth are present in the oral cavity. It ordinarily covers some six or seven years, beginning with eruption of the first permanent molar and ending with loss of the second deciduous molars.

During this state of flux, when teeth are coming, going, shifting, turning, and seeking a permanent state of balance, much may appear abnormal to the untrained eye which is in fact mere routine development. Although many malocclusions have their inception at this time, bites that are open may eventually close, malposed teeth may drift into alignment, and first permanent molars in cusp-to-cusp relationship may, with arrival of the bicuspids, interdigitate normally. It is a time when muscle forces, specifically those of mastication and deglutition, play a major role in developing the dentition. Yet it is a time when the therapist should be in close contact with the dentist, who is able to separate the routine from the incipient abnormality.

Dental Occlusion and Malocclusion

The neophyte, attempting to understand dental occlusion, may find the effort distressing. Many authors have described it, but have used varying terms to define a variety of concepts. Much will depend upon the dentist to whom you are speaking at the moment; s/he may incline toward a very narrow or quite broad interpretation.

We are safe in assuming that occlusion has to do with the bringing together of the upper and lower teeth. Since the upper teeth are immobile, we may take the next step and view this as a mandibular function. The end product should be the interdigitation of the cusps of the teeth in some manner.

The structure of the temporomandibular joint makes possible a wide range of motion other than a simple hingelike closure. Occlusion is thus not a fixed, static position; it is best thought of as an act, a motion, a verb; it is what occurs when the mandible moves from the resting position, where there is freeway space (vertical space between the cusps of the two arches) past the point of contact of the multidirectional inclined planes of the cusps—as many as eighteen inclined planes on a lower first molar—through the process of twisting and rotation that is brought to bear on the individual tooth when these inclined planes slide down each other, to a terminal position of interlocking cusps. If one wishes to indicate only the final phase, it is necessary to append the word "position." Occlusion and occlusal position are not synonymous terms. An illustration of the latter is seen in Figure 3-6.

It is well to differentiate also between occlusion and *bite*. The dentist who remarks that "his/her bite is good," is referring to the fact that the basic features of the patient's habitual occlusion are satisfactory, but is in no way implying that there are not individual teeth, or several of them, which are malpositioned in some way.

Also within this province, one concept that seems to be poorly understood by laymen is the distinction between "overbite" and "overjet." In

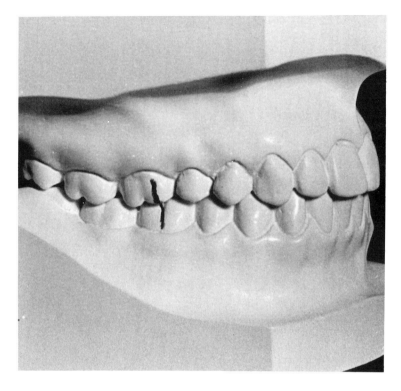

Figure 3-6. Normal centric occlusion of permanent teeth. All teeth are in contact, with normal molar relationship.

normal occlusion, the incisal edges of the mandibular teeth occlude against the lingual surface of the maxillary antagonists in such a way that approximately one-third of the mandibular crown is concealed, as shown in Figure 3-7. This overhang is sometimes called "vertical overbite" but more commonly is known simply as *overbite*. It is a strictly vertical dimension. If the incisors are in edge-to-edge relationship during habitual occlusion, there is zero overbite. When a space or vertical distance remains between the incisal or occlusal surfaces during habitual occlusion, the result is called *open bite.*

On the other hand, *overjet* is an anteroposterior dimension and implies the distance on the horizontal plane between the lingual surfaces of the upper incisors and the labial surfaces of the lower incisors.

Crossbite is seen in Figure 3-8 and refers to one or more teeth that are malposed buccally, lingually, or labially, with reference to the opposing tooth or teeth.

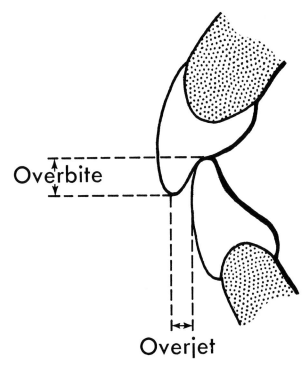

Figure 3-7. Distinction between overbite (a vertical dimension) and overjet (a horizontal dimension).

Malocclusion

Etiology. As the field of orthodontics progresses toward becoming a more exact science, it has become evident that more pervasive concepts of multiple causation must supplant older theories of simple cause-and-effect relationships. It is now generally accepted that there is strong interplay and interdependence of extrinsic and intrinsic factors on prenatal and postnatal influences in the emergence of malocclusion.

While tongue thrust may well prove to be one of the factors contributing to a given malocclusion, we would warn against placing unwarranted emphasis on this one aspect. Teaching a child to swallow correctly does not assure beautiful teeth or effortless orthodontic treatment. But it helps.

Classification of Malocclusion. The man who brought order out of chaos in the young field of orthodontics was Dr. Edward H. Angle (1855–1930). He inspired zealous adherents and bitter antagonists, but his intense efforts resulted in a systematization of knowledge formerly unknown. His name lives on in the name of a professional journal and a profes-

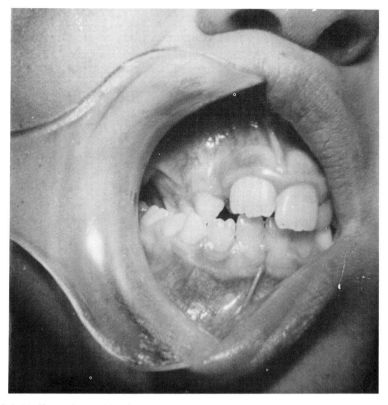

Figure 3-8. Unilateral crossbite. Mandibular molars are not contained within upper arch.

sional organization but is probably voiced most often in connection with the Angle system of classification of malocclusions. He did not have the benefit of modern research, and many of his ideas proved erroneous, but the Angle classification is used quite extensively today much as it was presented in 1907, despite more than eighty years of attacks on its adequacy. Angle was preoccupied with the position of the maxillary first permanent molar as the foundation on which to build, "the key of occlusion," and thus saw occlusion as a static horizontal relation of arch to arch, neglecting considerations on the lateral and vertical planes and the dynamics of dental function. The following is a restatement of the Angle classification (Angle, 1907, pp. 36–59), with Lischer's terms (1912, pp. 89–96) shown in parentheses:

Class I (neutroclusion). Those malocclusions characterized by normal mesiodistal relationship between the mandible and maxilla. The mesio-buccal cusp of the maxillary first permanent molar articulates in the mesiobuccal groove of the mandibular first permanent molar, as shown in Figure 3-9. Malocclusion occurs in the anterior segments; one or many teeth may be deflected from their normal course.

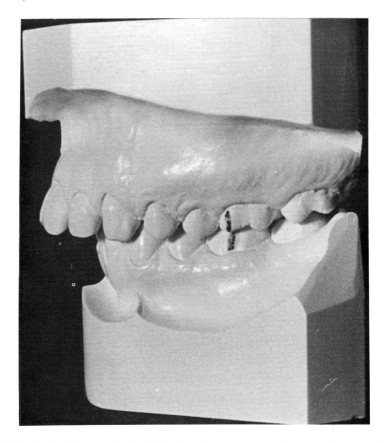

Figure 3-9. Angle Class I relationship. First permanent molars are in normal posture; malocclusion is confined to anterior teeth.

Class II (distoclusion). Those malocclusions in which there is retrusion of the mandible, the lower arch being distal from normal in its relation to the upper arch. The mesiobuccal groove of the mandibular first permanent molar articulates posteriorly to the mesiobuccal cusp of the maxillary first permanent molar.

> *Division 1.* Bilaterally distal, the maxillary incisors being typically in extreme labioversion (Fig. 3-10). Angle noted that these cases were primarily associated with mouth breathing.
> *Subdivision.* Unilateral distoclusion, molar relationship normal on one side of the dental arch.

Division 2. Bilateral distoclusion in which the maxillary central incisors are near normal or slightly retruded, whereas the maxillary lateral incisors have tipped labially and mesially (Fig. 3-11).

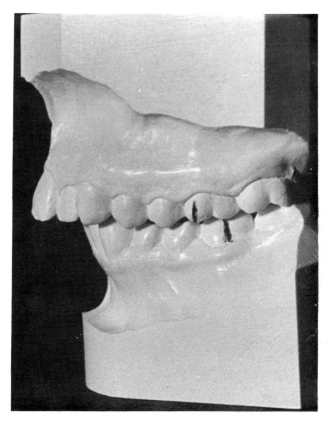

Figure 3-10. Angle Class II, Division 1, malocclusion. Entire mandibular arch is retruded, with upper incisors in labioversion.

Class III (mesioclusion). Those malocclusions in which there is protrusion of the mandible, a mesial relationship of mandible to maxilla. The mesiobuccal groove of the mandibular first permanent molar articulates anteriorly to the mesiobuccal cusp of the maxillary first molar (Fig. 3-12).

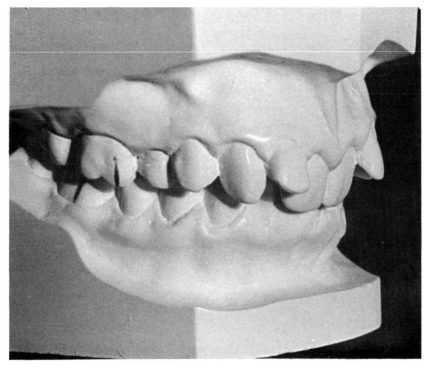

Figure 3-11. Angle Class II, Division 2, malocclusion. Note characteristic posture of upper incisors.

Some Dental Procedures

Treatment of the Young Child

Despite the fact that many schoolchildren now aspire to the status symbol of full bands, many millions of American children with malocclusion will never see an orthodontist. It has been estimated that over 50 percent of our population have dental irregularities; obviously, a much smaller percentage receives orthodontic treatment.

The American Dental Association has long urged the forestalling of problems before they occur, and it has laid this responsibility on every dentist to protect the dental interests of the majority of children. Dental service should be viewed in the progression: prevention, interception, correction. Time is the differentiating factor. If handled early, many dental defects can be avoided; if allowed to start, many can still be redirected before permanent damage ensures; if all else fails or if nothing has been attempted, only formal correction of the abnormalities remains.

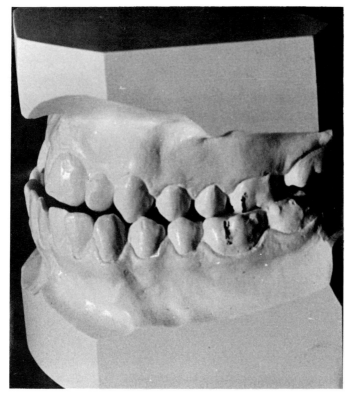

Figure 3-12. Angle Class III malocclusion. Entire mandibular arch is protruded. In extreme form (prognathism), mandibular arch engulfs the whole of maxillary arch.

Preventive Orthodontics

Preventive orthodontics should be the concern of the general dentist and pedodontist. Its aim is to eliminate or minimize the need for orthodontic attention at a later date.

The repair of caries is important; maintaining the primary teeth in good condition and position until they are replaced normally by the permanent teeth is a vital step in preventing malocclusion. The entire critical eruption cycle should be guided as necessary by close observance of the pattern of resorption of the deciduous roots and by the judicious use of space maintainers when early loss is sustained. Attention should also be given to gingivitis or other periodontal disease, tooth injury, ectopic eruption, supernumerary teeth, and other factors that influence the eruption of the permanent dentition.

Serial extraction of teeth, when indicated, is a preventive measure; it will be discussed separately, as will occlusal equilibration, which can be a major factor in both preventive and interceptive services.

The labial frenum may require careful evaluation by the dentist, for proper management can be most essential when diastema is truly the product of the frenum. The upper labial frenum normally attaches to the alveolar crest at birth but usually migrates in a superior direction as teeth and alveolar ridge develop; in some cases this migration is incomplete, and instead a body of tough, fibrous membrane develops between the maxillary central incisors, as seen in Figure 3-13. Surgical excision of this unyielding mass may allow closure of the diastema, although some dentists believe that it should not be done until the permanent canines have completed eruption. Anterior spacing may be a part of normal development, it may be hereditary, or it may be the product of abnormal function of the tongue. In those cases in which tongue pressure is maintaining the space, early frenum excision allows scar tissue to form at the site. The cicatricial mass is as resistant as the original frenum, and the result is therefore disappointing. If the tongue malfunction is corrected prior to the oral surgery, mesial drift or the eruption of the permanent second molars may quickly eliminate the space.

It is in the area of oral habits that we may yet make our greatest advances in preventive orthodontics, and it is here that the orofacial myologist may make an important contribution. Any effort that succeeds in providing the teeth with a more normal environment wherein to unfold would obviously increase their chances for *normal* development.

Interceptive Orthodontics

Rather than maintaining an occlusion that is still normal, the interceptive program attempts to restore normal occlusion in a mouth already going astray. Many of the procedures mentioned as preventive measures may be extended into use here. Certainly oral habits play a major role; elimination of thumb sucking, lip habits, cheek biting, tongue thrust, etc., is fundamental. Occlusal equilibration, extraction of supernumerary or ankylosed teeth, and education of parent and child concerning the importance of dental health, are all as essential here as in prevention. It has been reliably estimated that over one-fourth of all malocclusions can be intercepted given proper procedures at the proper time.

Interceptive procedures, however, generally imply a wider use of mechanical force. Some of the simple removable appliances described in the following chapter may be employed. A central incisor erupting in a

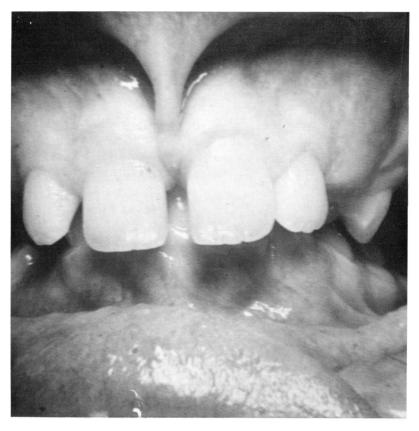

Figure 3-13. Abnormal maxillary labial frenum. Hypertrophied frenum has failed to migrate upward, thus maintaining diastema between central incisors.

crossbite position may be corrected in two or three weeks by the carefully directed use of a tongue depressor. Abnormal spacing in the anterior maxillary segment may be closed by various means, although indiscriminate measures can lead to serious problems. Lip exercises or the playing of a wind instrument may be prescribed to stimulate a flaccid upper lip, which in turn is allowing incisal protrusion. The propensity of the dentist for mechanical appliances has led to the use of a wide array of wire and plastic devices that exercise muscles, stop finger sucking, correct tongue thrust, and prevent mouth breathing.

Serial Extraction

The adjectival element in "serial extraction" is the operant word. This concept is not related to routine tooth extraction but rather is a carefully planned and precisely timed program involving removal of selected teeth over a period of four or five years, either in combination with mechanical treatment or, in ideal cases, as the sole procedure in preventing the development of malocclusion.

Ordinary tooth sacrifice during orthodontic treatment is not rare, for the removal of four bicuspids is commonplace. The critical task in any orthodontic program is not moving teeth but finding a place to put them. It seems that nature is generous in proportioning the teeth but does not squander a fraction of a centimeter on alveolar parking lots. Thus when the orthodontist adds up the width of all the teeth in an arch and then measures the total length of the arch itself, s/he usually finds a shortage of bone. No successful attempt has ever been made to increase permanently the available alveolar distance despite the fact that until 1940 most orthodontic treatment was aimed at "expanding the arch." Since the tooth-to-arch ratio must be balanced, an ear was finally turned to those who, for years, had been practicing and preaching a reduction in the number of teeth in the arch. Dr. Angle had promulgated an almost worshipful regard for each tooth and believed that every one was necessary for a normal dentition.

Now that the principle of essential tooth sacrifice has been accepted, programs first proposed 200 years ago have been perfected for orderly and optimum removal in physiological sequence. This is serial extraction. Although individual differences may demand a change in continuity, a frequently employed program starts at 5 or 6 years of age with removal of the four deciduous cuspids, thus allowing the eruption and optimal alignment of the permanent lateral incisors. At a later date, or at the same time if the patient has not been seen sufficiently early, the first deciduous molars are extracted in hopes of encouraging the eruption of the first bicuspids ahead of the cuspids. Since nature does not always cooperate, it occasionally becomes necessary to surgically remove the bud of the undeveloped first bicuspid, but this measure is avoided if possible. Most orthodontists prefer to wait until the first bicuspids begin to erupt, and then remove them, allowing the permanent cuspids to drop distally into the space created. In well-handled cases the patient finishes with normal occlusion, short only the expendable first bicuspids and having undergone no mechanical treatment. More often, some banding procedure is also necessary. The presence of an abnormal muscular force, such as a tongue thrust, jeopardizes any serial extraction program. However, when myofunctional therapy is provided shortly after eruption of the permanent lateral incisors, it serves as a strong adjunct to the extraction program, and the results can be most gratifying.

Preventive and interceptive measures influence incipient malocclusions or modify developmental factors that induce the problem. Given some of the etiological factors that are now recognized, malocclusions become inevitable. We will devote the next chapter, in its entirety, to the matter of corrective procedures, the final step in the series.

Items From Dental Specialties

Myofunctional disorders exert their influence upon almost every specialty in dentistry, although these forces are frequently overlooked or misinterpreted. We will not challenge the reader's retentive capacity by reciting the full catalog of such interrelationships. Instead, we offer specimen items from a number of areas, to lift the clinician's view beyond orthodontics to a few of the other pervasive ramifications that should be recognized.

Pedodontics. It is the pedodontist who should be the preventive and interceptive expert. S/he is often the first to see the patient, and at an age when the full range of options is still at his/her disposal.

Traditionally the dedication of the pedodontist has been primarily to the battle against tooth decay—repairing cavities and teaching proper prophylaxis. As the pedodontist becomes more cognizant of muscle dysfunction, the battlefield broadens.

Nevertheless, even if we limit our attention for the moment to a basic solicitude for tooth decay, we find myofunctional implications for the pedodontist. Dentists have frequently noted the detergent effect of saliva on oral hygiene. In normal deglutition, saliva is sucked around the teeth and through the gingival embrasures with a cleansing result. In abnormal function, this frequently repeated procedure is not only absent but in some respects is reversed, as when food particles are forced into the proximal spaces during a meal by the piston action of the tongue unmodified by a strong sucking component. Should it be proved that tongue thrusters are, in truth, more cavity-prone as a result of their dysfunction, pedodontists might well become responsive to this plight.

An extension of the foregoing is seen in the child with a problem of drooling. Emotional factors may be far more traumatic than dental aspects, for the social rejection which this child encounters may be almost constant. We have accepted into therapy a number of children at a somewhat earlier age than we might otherwise have liked, simply to ameliorate societal crises. The teacher seeking removal of a child from her classroom may have been only mildly exaggerating in her note to the parents complaining that the child "does all of her seatwork in a smelly pool of saliva." Although we cannot charge the pedodontist with responsibility for anticipating such occurrences, h/she should at least be aware of the possibilities so that h/she will be able to counsel the parents who are faced with this situation.

Periodontics. While there are other possible choices, we will present equilibration as our example from the field of periodontics.

Although the periodontist has, in the past, believed that he had less occasion to be aware of myofunctional disorders than some of the other dental specialists involved, he probably sees more of their deleterious effects than the orthodontist, who has made them a major concern. In part this may result from the age group that the periodontist treats; many patients are beyond the age when modification of muscle patterns seems feasible. Also, malocclusion is often viewed as a result of "premature" contact, the touching of certain features of the occlusal surfaces of the two dental arches in such a manner as to prevent normal intercuspation of the teeth. Yet tongue thrust is not always a suspected factor in the etiology of these interfering occlusal contacts.

Occlusal equilibration involves the detailed study of the occlusal contact of each tooth with its antagonist, the elimination of the premature or interfacing occlusal contacts mentioned above, the "wheel balancing" of the entire stomatognathic system, and through these a distribution of the load to all elements in equal proportion during mastication and deglutition. The *stomatognathic system* consists of the bones of the skull, plus the mandible, hyoid, clavicle, and sternum; the muscles and ligaments attached thereto; the dento-alveolar and temporomandibular joints; the vascular, lymphatic, and nerve supply systems; the soft tissues of the head; and the teeth. Note again that the tooth-alveolus articulation is treated as a joint.

The goal of equilibration is to assist nature in making compensatory adjustments throughout the stomatognathic system that will allow the forces dealt to and through the teeth to be distributed within normal limits to all components of the system. This prevents harmful stress on any given structure and provides for equal wear of all occlusal surfaces, thus smoothing the operation of the stomatognathic system and enhancing its integrity.

The actual process of equilibrating the occlusion may be quite intricate. Basically, it consists of making study models of both arches, making a record of the patient's bite on sheets of wax, identifying from these sheets any points where there is premature contact of cusps as the patient closes in occlusion, and then reshaping the teeth by grinding away the unwanted contact or by forming an inclined plane that will direct the bite into a normal position. When properly done, the forces of occlusion, working through the inclined planes of the teeth, can actually move certain teeth into a more functional position and restore harmony to the stomatognathic system.

When, on the other hand, the patient has even a mild anterior thrust, there can be unwelcome sequelae. Second molars, held out of occlusion by the interdental tongue tip, may start to supererupt, opening the bite still more. Repeated grinding on the occlusal surfaces may lead to destruction of the second molars. At that point, first molars may begin a similar eruptive pattern. We have seen several tongue-thrusting patients who have undergone equilibration, 30 to 50 years of age, with few teeth remaining distal to the bicuspids.

We may note in passing that such patients provide a great challenge for the orofacial myologist. With a reduced complement of posterior teeth comes reduced proprioception, posing an additional problem in retraining. When present in the mouth, these occlusal surfaces are the telegraphic keys that transmit signals through the periodontal membrane to the neuromuscular complex and thence to outlying areas of the stomatognathic system. When degeneration and destruction are permitted for any reason, it is unfortunate. When they are encouraged, even wholly unintentionally, by grinding teeth in a dysfunctional mouth, the consequences are still more deplorable.

Oral Surgery. The relevance of myofunction to the surgeon might seem obscure at first glance, and certainly there is a limited number of occasions when such considerations are germane. However, there are two instances of overlapping jurisdiction which are of such critical importance to the patient that their existence should be recognized and explained.

Our negative bias toward partial glossectomy as a solution to abnormalities of deglutition is beyond question. This procedure has been employed historically in many countries and is still suggested by some adherents. Its utilization emerged from the concept of tongue thrust as being predominantly a product of macroglossia, so that reduction of the bulk of the tongue should spontaneously effect normal function.

In cases of true macroglossia such a reduction might logically seem to have merit; we have had no personal contact with such cases and thus no opportunity to observe the results. The patients that we have seen post-operatively have all involved tongues whose original size may well have been within the normal range, and the results have been unfortunate. Basic patterns of muscle movement are little altered by the surgery—the thrust remains a thrust even though the blow is delivered by a lighter instrument.

In point of fact, the thousands of patients that comprise our combined total have provided very few cases of true macroglossia. Countless others have appeared so, on superficial examination, due to abnormally fronted function and lalling posture. The repositioning inherent in therapy renders a dramatic change in this deceptive illusion of enlargement.

On the other hand, we have been privileged to observe a few tongues such as that seen in Figure 3-14, A. This is a case of true macroglossia. The characteristic scalloping of the lingual margin is obvious; the imprint of each tooth is permanently embedded in the tissue. The dental structure enclosing this massive tongue is shown in Figure 3-14, B. Without the benefit of any type of orthodontic treatment, this 30-year-old occlusion is nearly perfect. It belongs to a speech pathologist whose articulation is flawless and who never personally required speech correction. It would appear that mere bulk is not the sinister influence that its reputation portrays, given compensatory function.

The second—and most frequently recurring—situation in which muscle dysfunction should concern the oral surgeon is that surrounding ostectomy. An entire new field of "surgical orthodontics" has emerged in recent years, devoted to the surgical removal of various segments of either jaw in order to normalize occlusion.

A precursor, a procedure by which to reduce the size of the truly prognathic mandible supporting a tongue of reasonably normal size and function, can produce gratifying results. The same can occur with the newer procedures entailing resection of maxilla or mandible, even with tongue thrusters—once function is habilitated and this contributing cause of the deformity removed. The aftermath can be quite opposite when no postoperative provision is made for myotherapy.

The premise is usually voiced that during the prolonged period after surgery when the mouth is wired shut, the tongue is *forced* to behave

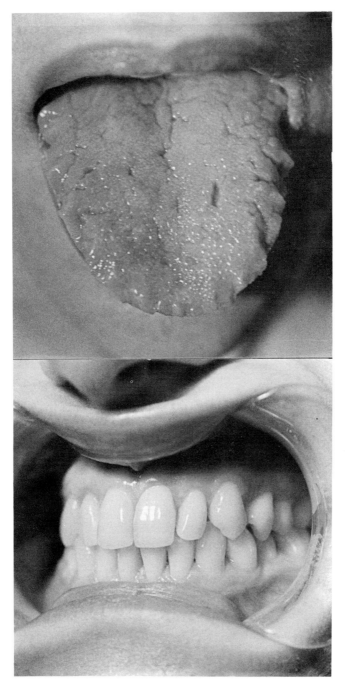

Figure 3-14. Patient with true macroglossia. A, Tongue shows imprint of every tooth. B, Natural untreated teeth showing no malocclusion.

properly and thereafter adjust itself to the altered environment. This idea harks back to the orthodontic assumption that correcting the occlusion eliminates tongue thrust. Experience indicates that both suppositions frequently prove misleading; nevertheless, the surgeon may routinely attribute relapse to noncooperation of the patient in not persevering in an ability never truly acquired by the patient. In our observation the contrast is thus striking between the patients for whom we have provided therapy prior to surgery and the resentful, angry patients who make their initial appearance postsurgically.

Prosthodontics. The prosthodontist is also faced with several manifestations of faulty muscle function. His/her most intense frustration may be reserved for the patient whose flailing tongue dislodges a carefully-crafted denture with each swallow, at times with a disconcerting clatter.

Such a patient is not necessarily of advanced age. Since myofunctional disorders contribute to early destruction of the teeth, the age at which dentures become necessary is accordingly lowered. A few patients in their twenties have been seen who were already wearing prosthetic substitutes. Several patients in their early thirties, faced with the imminent probability of having to wear dentures, have been seen in recent years. They completed therapy with varying degrees of success, but at least some of them were enabled to have their own teeth placed back in occlusion orthodontically and seemingly ward off dentures.

Since the anterior teeth are frequently the first affected, a fixed prosthesis in the incisal region is a common first step down the denture trail. The patient may then bend the prosthesis with incessant tongue pressure, and thereafter break the artificial teeth from the denture during incision. In some cases it has been necessary to construct the denture in the unattractive arch form of the original malocclusion, thus providing a "built-in tongue thrust," as an escape route for the tongue and to prevent further damage to the prosthesis. Such cases make it difficult to accept the phenomenal adaptability of the tongue that is postulated by those who rely on orthodontic treatment alone to correct swallowing patterns.

Conclusion

This has been a brief, broad, "shotgun" glance at the vast field of dentistry. Included have been only terms and concepts about which the clinician might reasonably be expected to converse intelligently in the process of dealing with problems of deglutition. The references listed after this chapter contain a great storehouse of information and are heartily recommended to those who wish to delve deeper.

The glossary provided immediately hereafter includes most of the essential terms. The clinician should have some recognition of each term. The majority of the items have been used or discussed in this chapter, some will appear in the chapter to follow, while the others seemed to require only definition.

GLOSSARY OF DENTAL TERMS

aberrant. Wandering or deviating from the usual or normal course.

abrasion. The wearing away of tooth substance by mechanical means.

abutment. A tooth used for the support or anchorage of a fixed or removable prosthesis.

alignment. The line of adjustment of the teeth.

alveolar process. The ridge projecting from the lower surface of the body of the maxilla or the upper surface of the mandible containing the alveoli of the teeth.

alveolar septum. The bony wall that separates individual alveoli.

alveolus; pl., alveoli. A tooth socket. A cavity in the jawbone that envelops the root of the tooth.

anatomical crown. The portion of the tooth that is covered by enamel.

anatomical landmark. A readily recognizable anatomical structure used as a point of reference in establishing the location of another structure or in determining certain measurements.

angle. A sharp bend formed by two borders or surfaces. A point. The angle of a tooth is the line or point where two or more surfaces meet.

ankyloglossia (tongue-tie). Partial or complete fusion of the tongue with the floor of the mouth or the alveolar crest; caused by lingual frenum being abnormally short or abnormally attached.

ankylosis. Abnormal immobility and consolidation of a joint. Stiffened; held by adhesions. An ankylosed tooth is fused to alveolar bone, with obliteration of the periodontal membrane.

anodontia. Total congenital lack of teeth, often combined with lack of sweat glands, persistence of fetal hair (lanugo), and defects of the nails. See also **oligodontia.**

antagonist. A tooth in one jaw that articulates with a tooth in the other jaw.

anterior tooth. Any one of the incisors or cuspids in either jaw.

apex; pl., apices. The pointed extremity of any conical part. The terminal end of the root of a tooth.

apical base. The basal bone portion of maxilla and mandible; that immediately adjoining portion upon which the teeth and alveolar process rest.

apical foramen. The opening of the pulp canal at the apex of the root of a tooth.

aplastic. Having imperfect development.

aptyalia. Deficiency or absence of saliva.

articulation (of teeth). The contact relationship of maxillary and mandibular teeth as they move against each other.

attrition. The wearing away of the incisal edges and occlusal surfaces of the teeth in the act of mastication or by the opposing teeth of the opposite jaw in the course of normal use.

axial surface. Any surface of a tooth that is parallel with its long axis. The labial, buccal, mesial, distal, and lingual surfaces are axial surfaces.

balanced occlusion. An ideal relationship of the mandibular and maxillary teeth to one another in centric position and throughout all the movements of the mandible.

balancing occlusion. The dynamic relationship of the mandibular and maxillary teeth to one another during the excursion of the mandible from balancing position to centric position.

balancing position. The static relationship of the mandibular and maxillary teeth to one another on one side of the dental arch when closure is made with mandible moved laterally to the opposite side.

basal bone. The bone of the maxilla and mandible, excepting the alveolar processes.

bicuspid. A tooth having two cusps or points. Man has eight bicuspids, also called premolars. They are situated between the cuspids and the molars, two on each side in both jaws. They are named from the median line distally as maxillary or mandibular first and second bicuspids.

bifid. Separated into two parts.

bifurcation. Division into branches. The division of a root into two parts. The division of a groove into two branches. The anatomic area where roots divide in a two-rooted tooth.

bruxism. Grinding of the teeth, especially during sleep. Also called stridor dentium.

buccal. Pertaining to the cheek. The buccal surface of a tooth is the surface next to the cheek.

calculus (tartar). A hard, mineralized deposit attached to the teeth.

canine. (1) Of, pertaining to, or like that which belongs to a dog. (2) The third tooth from the medial line. See also cuspid.

Carabelli cusp. The cusp located on the lingual surface of many maxillary first permanent molars. it is also known as the Carabelli tubercle and as the fifth cusp. (Named after Georg C. Carbelli, Vienna dentist (1787–1842).

caries. A localized progressive disintegration of a tooth, beginning with the solution of the enamel and followed by bacterial invasion; a "cavity."

cementoenamel junction. The line on the surface of a tooth that marks the meeting of the cementum and enamel. The cervical line.

cementum. The layer of bonelike tissue covering the root of a tooth. It differs in structure from ordinary bone in containing a greater number of Sharpey's fibers.

central fossa. The depressed area in the occlusal surface of the molars that surrounds the central pit.

central incisor. The first tooth on either side of the median line in either jaw. Also called the first incisor.

central lobe. The middle portion of enamel when the surface or part has three lobes.

centric occlusion. (1) The relationship of the teeth to each other when the jaws are closed so that the lingual cusps of the maxillary bicuspids and molars, and the buccal cusps of the mandibular bicuspids and molars, rest in the deepest parts of the sulci of the maxillary bicuspids and molars. (2) The relationship of the upper and lower teeth to one another when the jaws are complete closed and at rest.

cervical border. The extreme margin toward the root of any axial surface of the anatomical crown of a tooth. It is located at the cervical line.

cervical ledge. The slight elevation of enamel around the periphery of the crown immediately above the cervical line.

cervical line. (1) The line of the anatomical neck of the tooth; to be distinguished from the gingival line. (2) The line around the surface of a tooth where the enamel and cementum meet.

cervix. The neck or any neckline part. The cervix of a tooth is the portion of the tooth surface adjacent to the junction of the crown and root.

cicatrix. A scar left by a healed wound.

cingulum; pl., cingula. The lingual lobe of anterior teeth which is located mostly in the cervical third of the lingual surface.

clicking (TMJ articulation). A snapping or cracking noise evident on excursion of the mandibular condyle. See also crepitus.

clinical crown. (1) The portion of the tooth that projects from the tissues in which the root is fixed. (2) The portion of the tooth that is visible in the mouth.

comminution. The act of breaking, or the condition of being broken into small fragments.

condyle. The rounded eminence at the articular end of a bone. That portion of the mandible that articulates with the temporal bone of the skull to form the temporomandibular joint.

contact area. The portion on the surface of a tooth that touches the adjacent tooth in the same arch.

crepitus. (1) A grating sound heard on movement of ends of a broken bone. (2) The cracking sound emitted by a dysfunctioning temporo-mandibular joint.

crypt. A follicle or tubule; a small glandular sac or pit.

curet (curette). An instrument having a sharp, spoon-shaped blade, used for debridement of periodontal pocket, tooth root, and bone.

curettement. Scraping or cleaning the walls of a cavity or surface by means of a curet.

cusp. A pronounced elevation or point on the crown of a tooth.

cuspid. The third tooth from the median line, lying between the lateral incisor and the first bicuspid, the incisal edge of cuspids is raised to form a single point or cusp. There are four cuspids in all. They are named maxillary right and left and mandibular right and left cuspids. Also called canine.

cutting edge. Same as incisal edge.

debridement. Slitting a constricting band of tissue, the surgical removal of lacerated, devitalized, or contaminated tissue.

deciduous tooth. One of the teeth of the first dentition, so called because they are shed to give place to the permanent teeth. Also called temporary or milk teeth.

dental dysplasia. Abnormal development of bone, resulting in insufficient space to accommodate all teeth.

dentin, dentine. The hard tissue that forms the main body of the tooth. It surrounds the pulp and is covered by the enamel and cementum.

dentinocemental junction. The line of meeting of the dentin and enamel.

dentition. The kind, number, and arrangement of the teeth.

diastema; pl., diastemata. A space between two teeth, commonly between the central incisors. Also called **trema.**

distal. Away from the medial line following the curve of the dental arch.

dorsum. (1) The back or posterior surface of any organ or part. (2) The upper surface and back of the tongue.

dysphagia. Inability or difficulty in swallowing; may result from hysteria, paralysis, muscle spasm, narrowing of pharynx or esophagus, etc.

ectopic eruption. In an abnormal position; a tooth erupted out of its normal sequence in the dental arch.

edentulous. Absence of teeth due to loss, as contrasted to anodontia, in which teeth never existed. Edentulous space: site of tooth loss either through trauma, extraction, or natural exfoliation of deciduous tooth.

embrasure. An opening with sloping sides; the sloping space adjacent to the contact.

> **buccal e.** The embrasure opening from the contact toward the cheek in the posterior teeth.
>
> **gingival e.** The embrasure opening from the contact toward the alveolar process. The interproximal space.
>
> **incisal e.** The embrasure opening from the contact toward the incisal edges of anterior teeth.
>
> **labial e.** The embrasure opening from the contact toward the lips in anterior teeth.
>
> **lingual e.** The embrasure opening from the contact toward the tongue.
>
> **occlusal e.** The embrasure opening occlusally from the contact in posterior teeth.

enamel. The hard, mineralized tissue that cover the dentin of the crown of a tooth.

endodontics. The specialty of dental science concerned with the diagnosis and treatment of diseases of the dental pulp.

endodontium. The dental pulp.

endogenous. Growing from within; developing or originating within the organism, or arising from causes within the organism.

epithelium. The epidermis of the skin; the surface layer of mucous membranes, consisting of one or more layers of cells varying in form and arrangement.

erosion. The loss of tooth substance due to a combination of chemical action and abrasion.

eruption. The emergence of a tooth through the soft tissues to appear in the oral cavity.

exogenous. Originating or deriving from outside the organism; being produced or growing from without.

extrusion. The hypereruption or migration of a tooth out of its normal plane of occlusion.

facet. A small abraded spot on a tooth.

facial surface. The surface of a tooth that is next to the lip or cheek; the vestibular or outer surface.

fissure. A fault in the surface of a tooth caused by imperfect joining of the enamel of the different lobes. Fissures occur along the lines of developmental grooves.

fossa; pl., fossae. A round or angular depression in the surface of a tooth. Fossae occur mostly in the lingual surface of incisors and the occlusal surface of posterior teeth.

freeway space. The space between maxillary and mandibular antagonist teeth when the mandible is suspended in postural rest position.

frenulum; pl., frenula. A small frenum. Sometimes applied to lingual frenum.

frenum; pl. frena. A fold of mucous membrane that serves to check the movement of a part or organ.

> **lingual f.** Fold along midline of inferior surface of tongue extending to floor of mouth.
>
> **labial f.** Folds at the midline that attach the upper and lower lip to alveolar tissue.

gingiva. The gum; the fibrous tissue covered by mucous membrane that covers the alveolar processes of the jaws and surrounds the necks of the teeth.

gingival line. The line of contact of the extreme border of the gingiva to the tooth; to be distinguished from the cervical line.

gingival papilla. The part of the gingiva that lies in the gingival embrasure.

gingival sulcus. The space that develops in the soft tissues surrounding the tooth, bounded by the tooth surface on one side and the epithelial lining of the gingiva on the other.

gingivally. A direction from any part of the tooth toward the gingival line.

gnathic. Pertaining to or affecting the jaw or cheek.

gnathology. (1) The science of the masticatory system, including physiology, functional disturbances, and treatment. (2) A specialized field of dentistry concerned primarily with positioning the teeth in healthy relationship with the temporomandibular joint; also called orthognathics.

groove. A linear depression in the surface of the tooth.

hyperplasia. The abnormal multiplication or increase in the number of cells in a tissue; an increase in size of a tissue or organ resulting from proliferation of cells.

hypertrophy. The enlargement or overgrowth of an organ or structure due to an increase in size of its constituent cells, but not resulting from an increase in the number of cells.

hypoplasia. Defective or incomplete development.

iatrogenic. Any adverse condition in a patient occurring as the result of treatment; a detrimental condition induced or caused by a doctor.

idiopathic. Of unknown causation.

incisal edge. The sharp angle formed by the union of labial and lingual surfaces of anterior teeth. The cutting edge of the anterior teeth.

incisal papilla. An oval or pear-shaped nipplelike prominence of the gingiva immediately behind the upper central incisors. Also called **palatine papilla.**

incisor. Any one of the four front teeth of either jaw.

inclined plane. A sloping area found on the occlusal surfaces of bicuspids and molars. It is bounded by the primary grooves and the crests of the ridges. Each normal cusp has two inclined planes named for the direction in which they face, that is, the lingual cusps have mesiobuccal and distobuccal inclined planes, and the buccal cusps have mesiolingual and distolingual inclined planes.

intercuspation. The cusp-to-fossa relationship of the maxillary and mandibular posterior teeth to each other.

interdigitation. The interlocking or fitting of opposing parts, as the cusps of the maxillary and mandibular teeth; intercuspation.

interproximate space. The V-shaped space between the proximal surfaces of adjoining teeth; it extends from the contact to the crest of the alveolar process.

keratin. A protein that forms the basis of hair, nails, and any horny tissue, including the organic matrix of tooth enamel.

labial surface. The surface of an anterior tooth that lies closest to the lips.

lamina propria. Alveolar bone proper, or cribriform plate. It lines the inner surface of the alveolus and offers attachment for the fibers of the periodontal membrane.

lateral incisor. The second tooth from the medial line on each side in either jaw. Also called the **second incisor.**

leukoplakia (smoker's tongue). Formation of white thickened patches on the mucous membrane of the tongue or cheek. These cannot be rubbed off, show a tendency to fissure, and may become malignant.

lingual surface. The tooth surface that is next to the tongue.

lobe. One of the main morphological divisions of the crown of a tooth.

long axis. An imaginary line passing lengthwise through the center of the tooth.

luxation. Dislocation of a joint, as the temporomandibular articulation, or displacement of organs.

malar. Pertaining to or affecting the cheek.

malocclusion. Imperfect or irregular position of the teeth.

mamelon. One of the three rounded prominences on the incisal edge of anterior teeth when they first erupt.

mesial. Toward the median line following the curve of the dental arch.

molar. One of the large grinding teeth of which there are three on either side in both jaws. They are situated distal to the bicuspids and named from before backward as maxillary or mandibular first, second, and third molars. The first molar is also called the **six-year** molar; the second molar, the **twelve-year** molar; and the third molar, the **wisdom tooth.**

occlusal surface. The surface of a bicuspid or molar that makes contact with a tooth of the opposite jaw when the mouth is closed.

occlusion. The contact of the teeth of both jaws when closed or during those excursive movements of the mandible that are essential to the function of mastication.

oligodontia. Congenital absence of one or a few teeth.

operculum. (1) Any covering. (2) The hood or flap of mucosa over an unerupted or partially erupted tooth.

orthodontics. The profession or science of straightening teeth.

papilla. Any small, nipple-shaped elevation.

> **incisive p.** The elevation of soft tissue covering the foramen of the incisive canal; crosses upper gingiva along midline behind maxillary central incisors.

> **interdental p.** Gingiva filling the interproximal spaces between adjacent teeth.

> **lingual p.** Any one of the tiny eminences covering anterior two thirds of tongue, including circumvallate, filiform, fungiform, and conical papillae.

pedodontics. Specialized care of children's teeth.

periodontal membrane. The fibrous tissue that is attached to the cementum of the tooth and to the surrounding structures.

periodontics. Phase of dentistry dealing with treatment of diseases of the tissues around the teeth.

periodontium. The investing and supporting tissues surrounding the tooth—the periodontal membrane, the gingiva, and the alveolar bone.

periosteum. Tissue that covers the external surface of a bone.

pit. A sharp pointed depression in the enamel.

posterior tooth. One of the teeth situated distal to the cuspids. Bicuspids and molars are posterior teeth.

primary groove. A sharp V-shaped groove that is a constant and developmental part of the tooth. Marks the union of the lobes.

prosthodontics. The branch of dentistry pertaining to the replacement of missing teeth by artificial devices, whether with dentures, or fixed or removable bridges.

proximal surface. One of the surfaces of a tooth, either mesial or distal, that lies next to an adjacent tooth.

pulp. The soft tissue containing blood vessels and nerve tissue occupying the central cavity of a tooth.

pulp canal. The part of a pulp cavity that traverses the root of a tooth.

pulp cavity. The entire central cavity in a tooth; it contains the dental pulp.

pulp chamber. The enlarged portion of the pulp cavity, which lies mostly in the central portion of the crown.

resorption. The gradual loss of the tooth structure or of bone resulting from an altered biochemical state in a localized area.

ridge. A long, elevated portion of the tooth surface.

root. The portion of a tooth that is covered with cementum.

root canal. Same as **pulp canal.**

ruga; pl., rugae. Irregular, sometimes branching ridges across the hard palate, radiating from the incisal papilla and the palatine raphe.

secondary groove. A groove of lesser importance. Secondary grooves differ from primary grooves in that they are usually rounded, or U-shaped, at the bottom, and they do not mark the boundaries of the lobes.

septum. A dividing wall or partition. One of the thin plates of bone separating the alveoli of the jaw.

stomatognathic. Pertaining to the unified structure and function of mouth and jaw with all appurtenant tissues and organs as a cohesive system.

subluxation. Incomplete or partial dislocation.

succedaneous tooth. Permanent tooth that succeeds or takes the place of a corresponding deciduous tooth.

sulcus; pl., sulci. A well-defined, long-shaped depression in the surface of a tooth, the inclines of which meet at an angle.

supernumerary. Exceeding the regular number. An extra tooth, often peg-shaped.

supplemental lobe. An additional lobe. A lobe that is not usually associated with the typical form of a tooth.

trismus. Inability to open the mouth due to spasms of the muscles of mastication.

trunk. The main body of the root of a multiple-rooted tooth. That portion of the root from the cervical line to the division of the root.

tubercle. A small, rounded, or pointed elevation of enamel. Tubercles occur frequently on the cingula of anterior teeth and occasionally on various parts of other teeth.

working occlusion. The dynamic relationship of the mandibular and maxillary teeth to one another during the excursion of the mandible from working position to centric position.

working position. The static relationship of the mandibular and maxillary teeth to one another on one side of the dental arch when the mandible is moved laterally to that side.

REFERENCES

Angle, E. H. (1907). *Malocclusion of the teeth* (7th ed.) Philadelphia, PA: S. S. White Dental Manufacturing Co.

Arey, L. B. (1954). *Developmental anatomy,* (6th ed.) Philadelphia, PA: W. B. Saunders.

Best, C. H., & Taylor, N. B. (1961). *The physiological basis of medical practice,* (7th ed.) Baltimore, MD: Williams & Wilkins.

Burlington Orthodontic Research Center [Progress report series no. 6] (1960–1961). Toronto, Can.: University of Toronto.

Case, C. S. (1921). *Dental orthopedia.* Chicago, IL.: C. S. Case Co.

Graber, T. M. (1966). *Orthodontics: principles and practice,* (2nd ed.). Philadelphia, PA: W. B. Saunders.

Langley, L. L., & Cheraskin, E. (1956). *The physiological foundation of dental practice.* St. Louis, MO: C. V. Mosby.

Lischer, B. E. (1912). *Principles and methods of orthodontics,* Philadelphia, PA: Lea & Febiger.

Moyers, R. E. (1973). *Handbook of orthodontics.* Chicago, IL: Year Book.

Salzmann, J. A. (1957). *Orthodontics: principles and prevention.* Philadelphia, PA: J. B. Lippincott.

Shore, N. A. (1959). *Occlusal equilibration and temporomandibular joint dysfunction.* Philadelphia, PA: J. B. Lippincott.

Sicher, H., & DuBrul, E. L. (1975). *Oral anatomy* (6th ed.). St. Louis, MO: C. V. Mosby.

Strang, R. H. W., & Thompson, W. M. (1958). *Textbook of orthodontia,* (4th ed.). Philadelphia, PA: Lea & Febiger.

Zeisz, R. C., & Nuckolls, J. (1949). *Dental anatomy.* St. Louis, MO: C. V. Mosby.

CHAPTER 4

ORTHODONTIC CONCEPTS AND PROCEDURES

Introduction

This chapter is offered for the reader who is without dental background, especially of an orthodontic nature. It is a blend of many rationales and dogmas; it may fit poorly the routine of a given working orthodontist; in the process of synthesizing we lose individual thought and opinion, levelling the best and worst into a general average. Since we are not undertaking to instruct dentists or to describe a specific orthodontic philosophy, perhaps we may be forgiven this glimpse through the eyes of a lay observer into the world of the orthodontist.

The orofacial myologist is strongly urged to pay a series of visits to one or more orthodontists, watch them work, become familiar with some of their procedures—what makes a headgear work, how bands are applied, and how force is transmitted to the teeth—and learn the nomenclature of some of the materials and appliances.

General agreement has been reached that both the profession and the treatment will be known as orthodontics. Do not speak of "braces," for this term refers either to prostheses for arms and legs or to antediluvian dentistry, when prefabricated monolithic devices were locked onto the teeth; use instead such terms as "bands" or "appliance," or other specific appellation.

Get acquainted with plaster study models and realize that they are not quick impressions of teeth but are carefully produced re-creations of the dental arches exactingly oriented to reflect either the Frankfurt or occlusal plane. When stood upright on their posterior surface, with incisors uppermost and the two halves approximated, they form a precise duplicate of the patient's occlusion as of the date taken. Models, infrequently known as "casts," are precious to the orthodontist, representing time, effort, and money, and cannot be replaced at a later time when the patient's occlusion is no longer identical. They chip and break easily. The quality of interpersonal relations may hinge on careful handing of models.

Appliances

As in speech therapy, the attitude and skill of the operator is often of far greater importance than the method used. Nevertheless, there are several schools of thought in orthodontics, all aimed at the same result but using different techniques and appliances. It is advisable to learn the philosophy of the dentists with whom you associate.

Most appliances were first invented by individual dentists, many of whom then patented the devices. Thus many appliances and systems came to be known simply by the name of their creator. Appliances fall into two general categories, removable and fixed.

Removable Appliances

Originally, all appliances were removable and in effect were dentures with built-in springs, prongs, hooks, pressure screws, etc. They have been traditionally the method of choice in Europe, there they were developed to a high degree. In the United States, where attention has been focused on fixed appliances, removable devices have been used primarily in mild, uncomplicated conditions, or interceptive or preventive orthodontics, or as retaining devices following full correction with a fixed appliance. The general practitioner, the pedodontist, and other nonspecialists in orthodontic treatment often feel more comfortable with removable appliances; they are usually more simple to construct and can be fabricated at less expense to the patient. In recent years, however, an increasing number of orthodontists in the United States have been using removable appliances, either as primary treatment devices, or in conjunction with fixed appliances. European dentists in turn, are now using fixed appliances to upgrade their treatment results. Graber and Neumann (1984) strongly recommend that orthodontists be trained in the use of removable *and* fixed appliances.

> A considerable proportion of orthodontic treatments can be carried out with simple appliances or as the first phase of treatment when functional appliances can be used in the interval after the emergence of incisors and before the emergence of canines and premolars. When properly timed, this period can be used to great advantage, and, if necessary, the initial treatment with functional appliances can be followed by fixed appliance therapy to normalize tooth position and occlusion through controlled tooth movements (p. 4).

Graber and Neumann (1984) assert that, properly used, removable appliances can correct many malocclusions and considerably improve others. Used in the mixed dentition stage, they can greatly simplify therapy in numbers of cases, without delaying or complicating further treatment in the permanent dentition if it is needed.

> The potential and very useful combination of fixed and removable appliances, and of the use of extraoral force with removable appliances, has hardly been explored yet. Once the profession has grasped the potential and the possibilities, the lingering doubts about the usefulness of removable appliance will soon be dispelled (p. 9).

There are two broad categories of removable appliances: (1) those that attempt to stimulate muscle activity which can be harnessed by the appliance and directed into the service of moving teeth; and (2) those that rely on the appliance itself or some attachment thereto as the means to generate the force required in tooth movement. This sorting arrangement is not entirely satisfactory, since it provides no pigeon-hole for the appliances intended to maintain static position and prevent tooth movement. For ease of discussion, therefore, we will divide them instead into two different classes, loose and attached. By and large, loose appliances include those in type 1 above, while attached removables take in those of type 2 plus the holding devices. There is minor overlapping.

Loose Removable Appliances. Loose removable appliances include, among others, the oral shield, the Andresen appliance, the Bimler appliance, the Frankel appliance, the positioner, and the Bionator.

An oral shield is simply a thin sheet of acrylic that is placed between the anterior teeth and the lips (the oral vestibule) and that extends distally to the maxillary second molars. It is accurately fitted to the individual mouth and is used to correct mild labioversion of maxillary incisors, close spaces in this region, stimulate lip action, and correct tongue thrust.

The Andresen appliance (Fig. 4-1) is also known as an "activator," the "Norwegian system," or commonly as a "monobloc." It is also acrylic, and lines the mouth in one continuous body lingual to the dental arches, conforming to the contours of the lingual surfaces. It frequently has a shelf that fits between the upper and lower teeth, with a replica of the occlusal surface of each tooth molded into the acrylic; however, these depressions are placed in a more advantageous position than in the existing bite. The appliance is quite loose fitting and falls if not held in position by the tongue or by closing the teeth. The aim is to stimulate muscle action, setting up new reflexes such as closing the teeth and lifting the tongue during deglutition, while guiding the mandible forward.

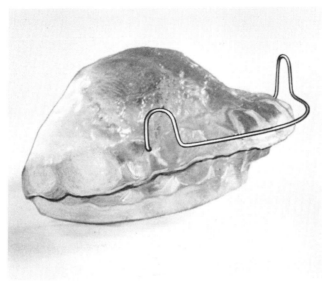

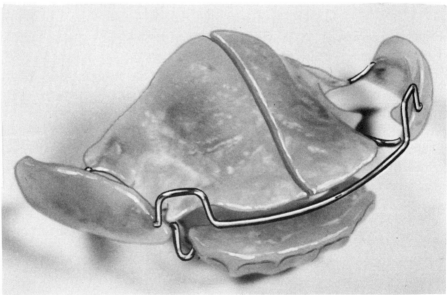

Figure 4-1. Andresen appliances. A, Conventional monobloc, designed to contain tongue. B, Activator with modifying attachments. (Courtesy Rocky Mountain/Orthodontics.)

The activator is most useful during the mixed and early permanent dentition period. It can partially or totally correct Class II, Class III, and open bite malocclusions. Graber and Neumann (1984) advise against its use in malocclusions involving crowding, and, with rare exceptions, for malocclusions requiring extractions. It is not effective, generally, in moving individual teeth.

The Bimler appliance, as seen in Figure 4-2, is composed basically of wrought wire with acrylic wings flaring up lingual to the buccal teeth and connected by a U-shaped wire called a Coffin spring. The molded continuous wire is adapted to the labial surface of the upper arch and may have additional springs attached in a given case. The anchorage obtained in the buccal segments permits movement of the anterior teeth, both upper and lower.

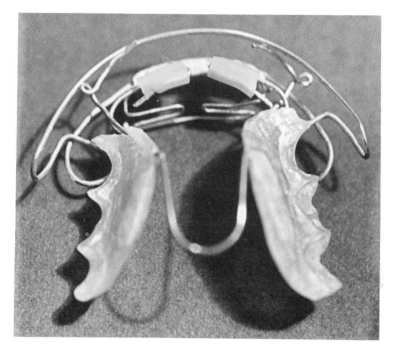

Figure 4-2. Bimler appliance, with acrylic palatal wings connected by Coffin spring.

The Bimler combines some of the principles of the labiolingual (fixed) arch technique with features of the Andresen (removable) activator, to produce an appliance that is designed for use in the early mixed dentition stage. It attempts to recruit both the growth and the orofacial muscle forces of the patient to the service of repositioning teeth, guiding growth into favorable patterns, and improving oral behavior. The Bimler consists of a labial arch wire in the upper arch and a lingual arch wire in the lower arch. These wires are not fixed to the teeth, but to each other. Graber and Neumann (1984) warn that, whereas certain types of cases show rapid, favorable results, others end up with unsatisfactory results. The Bimler may be used in Class II, divisions 1 and 2, and another type for Class III.

The Frankel appliance may take many strange forms. It is also constructed mainly of an elaborate wire framework, but is primarily tissue-borne rather than tooth-borne; in fact, it often does not actually touch the teeth. In the molar region it is equipped with acrylic pads that hold the buccal wall away from the dental arch and support the distal ends of the wire framework. The acrylic may extend between the upper and lower buccal teeth, as in the monobloc, or be augmented by another acrylic "lip plumper" behind the lower lip to reduce mentalis muscle activity. As with the Bimler and several other removables, the concept behind the Frankel "function regulator" is to change existing muscle pressures in such a way that desirable tooth movement and growth responses will result. The Frankel can be used to increase sagittal, vertical, and transverse intraoral space, to move the mandible forward, and to aid in the formation of proper tongue and lip postures and movements.

A *positioner* is made of rubber or flexible plastic, and as one unit it surrounds the crowns of all teeth in both arches. Such an appliance is seen in Figure 4-3. It is used as the final step of treatment by some orthodontists. When the fixed appliance is removed, impressions are taken, the teeth are cut off these models and reset into an ideal relationship, and the positioner then duplicates this ideal. It is worn principally at night, and the teeth are gently guided into their ultimate positions.

The Bionator. Another loose, removable appliance is one developed by Balters, called the Bionator. It is less bulky than the activator, and relies principally upon lingual activity to perform its functions. Balters contended that the tongue, operating in an oral space inadequate for its proper function, exerts harmful influences upon the teeth and upsets the desired equilibrium between lingual and labio-buccal forces. The purpose of the use of the Bionator is to correct mandibular retrusion and associated malpositioning of the tongue, and to correct intrusive tongue, lip, and cheek patterns. It assists in maintaining a closed-lip resting posture, guides the tongue into contact with the soft palate rather than against the anterior teeth, and enlarges the oral cavity. Graber and Neumann (1984) caution that success is often only partial, due to difficulties in securing necessary patient cooperation and to growth factors difficult to predict and control. When its use is successful, the occlusal

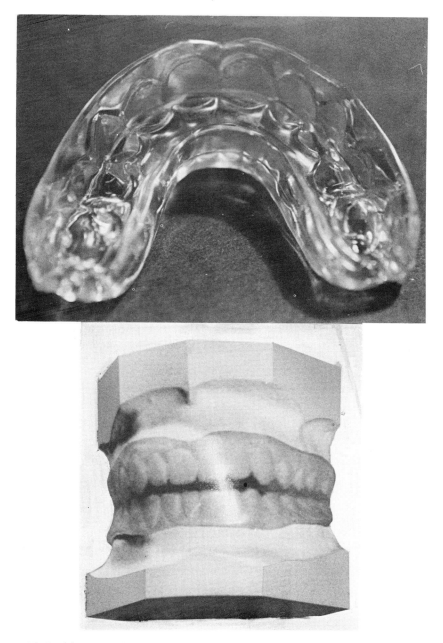

Figure 4-3. Positioner made of flexible plastic. A, View of positioner showing imprint of every tooth. B, Positioner in place. (Courtesy Rocky Mountain/Orthodontics.)

plane is corrected, the mandible is elongated, and the incisors are brought into normal relationships.

Attached Removable Appliances. Some of the more common types of attached removable appliances are the bite plane (or bite plate), the space maintainer, the stabilizing plate, the retainer (with variations), extraoral appliances, and the entire system of orthodontic treatment based on the Crozat appliance.

A bite plane is a palatal structure made almost entirely of acrylic and dependent on the mucosa and the teeth themselves for anchorage. It may have springs attached to stabilize it in the molar area, or hooks by which a very light elastic may be held around the upper incisors, or in some cases it may overlay the entire dental arch. Its main feature is a smooth sloping shelf against which selected teeth occlude, tipping them into a more desirable position. If the shelf is made level, such an appliance becomes known as a bite plate and may be used to open the bite; for example, if only mandibular incisors meet the plate, molars are held out of occlusion and thus encouraged to erupt further.

A space maintainer (Fig. 4-4), as the name implies, is used to maintain the space where a deciduous tooth or teeth have been lost, pending the arrival of permanent teeth. The acrylic covers the lingual surface of both the teeth and the mucosa, and extends into the edentulous space or spaces. It also prevents overeruption of teeth in antagonist position to the space by providing a surface on which they can occlude.

A stabilizing plate is an acrylic plate somewhat similar to the body of a space maintainer, except that it is molded precisely to the contour of the lingual surface of the arch (often mandibular) and has metal shafts that fit into tubes welded to molar bands. Its purpose is the positive maintenance of the molars in a desired position. It is little used, and then only in the more difficult cases.

The Hawley retainer is used to maintain the new positions of teeth at the conclusion of active treatment with a fixed appliance. It consists of an acrylic body accurately fitted to the lingual surfaces of the teeth, the maxillary version often being a replica of the entire palatal arch, rather than the partial arch shown in Figure 4-5. A variety of wires are incorporated, depending on the program that has been carried out or the problem to be solved, the wire fitting around the labial surface. Variations of the Hawley, which therefore should not be referred to as Hawleys, have been endless. Dentists have found the principle invaluable in a number of early or relatively minor problems. The labial wire can be activated to produce stress against certain malposed teeth, a light

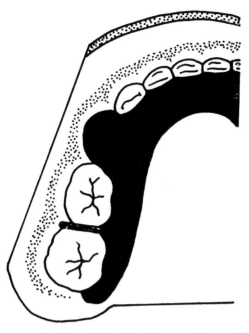

Figure 4-4. Acrylic space maintainer. Wire clasp between first and second molars holds plate in place.

elastic can replace the wire to close anterior spacing or reduce minor protrusion, space maintainers may be incorporated, etc.

Extraoral appliances were once described as "skullcap therapy," from their origin in the last century when some fairly weird harness contraptions were placed over the head and attached to the teeth. Today such appliances bear the more respectable title of "headgear" or "neckgear", and are in widespread use. The type shown in Figure 4-6, A, is designed for use with a full-banded appliance on the upper arch, the metal hooks of the headgear fit into special brackets that are welded to the bands. The cloth or plastic straps may go around the head in any of several arrangements and supply external anchorage. Although tooth movement can thus be actuated at an early age, the basic idea is to take advantage of growth and development, restricting and retarding the downward and forward growth of the alveolodental portion of the maxilla, while the mandibular bone continues its physiological growth pattern and "catches up" with the maxilla. The reverse is also possible in Class III cases, by using a "chin cup" on the outer surface of the chin rather than an intraoral attachment.

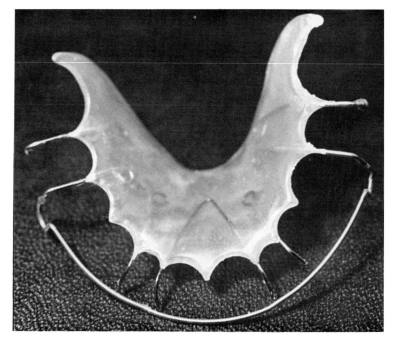

Figure 4-5. Hawley type of retainer. Acrylic palate supplies anchorage for labial wire.

A cervical strap is pictured in Figure 4-6, B. More commonly called a "neck band" and more commonly used than a headgear, it replaces the head straps with a single band, usually elasticized, which goes around the neck. A "face-bow" of steel wire is welded to the front of a heavy labial arch wire in one of various positions; the face-bow extends from between the lips around the outside of the cheeks, ending in a loop that attaches to the cervical strap. A pair of maxillary molars are fitted with bands bearing tubes on the buccal side to receive the ends of the arch wire. By modifying the arch wire, the face-bow, or their union, the dentist can produce a range of forces. This appliance is less cumbersome and more comfortable, but it provides less control of forces when teeth are being moved.

The Crozat appliance has won many fiercely loyal and vocal adherents, and the dentist so dedicated is often known as a "Crozat man." He is most frequently a pedodontist or general dentist, and usually belongs to a study club composed of other Crozat proponents for the purpose of sharing ideas and experiences, for this deceptively simply appliance can be most fickle in the unskilled hands of one who has not delved deeply into its vagaries. Few orthodontists use the Crozat, for their training

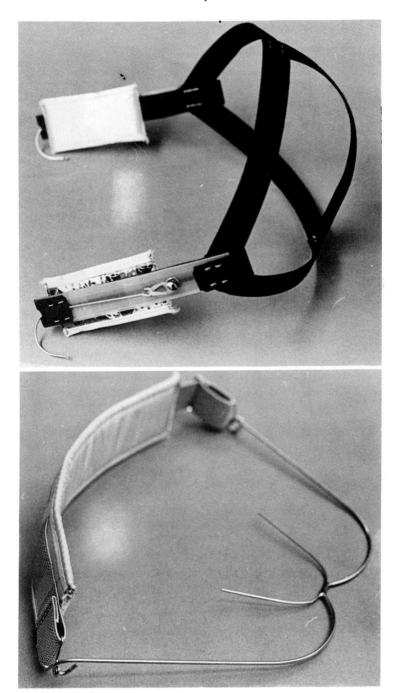

Figure 4-6. Two types of headgear. A, Anchorage is supplied by cranial straps fitted over patient's head. B, The "neck band," in which anchorage is achieved by cervical strap worn around patient's neck.

is otherwise; few dental schools offer instruction in this procedure. Nevertheless, some dramatic successes can be shown by the true expert to result from the use of the very light forces of this nearly invisible appliance.

As with most appliances, individual dentists invent personal variations, so that no two appear quite the same—partly due also to the unique needs of the patient. A few varieties are depicted in Figure 4-7; all have features in common. The most distinctive characteristic is the close-fitting but uncemented molar bands that hold the appliance in place. The wire body, usually of precious alloy, is generally fitted to the lingual of the mandibular arch, while a high labial arch wire typifies the upper appliance. From this maxillary arch wire, separate prongs often extend down to the labial surface of each tooth, the placement on the tooth and the tension of the individual fingers regulating tooth movement. Beyond these basic details, the manifestations of the appliance become too complicated to describe in a brief space.

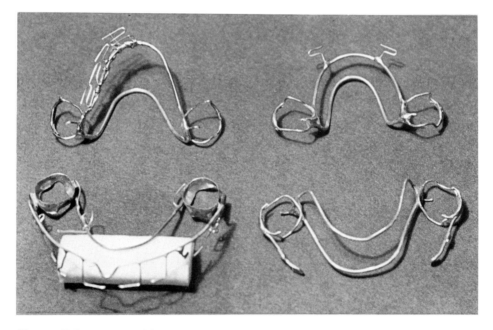

Figure 4-7. Some types of Crozat appliances. A, Correcting unilateral imbalance. B, Labial hooks to receive elastics. C, Propped on cotton roll to display finger springs, which may take various designs. D, Basic construction.

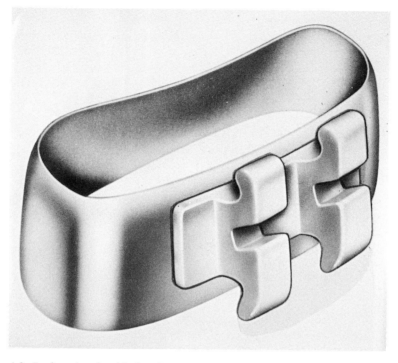

Figure 4-8. Incisor band with bracket welded in place. That pictured here is called "Siamese" bracket. (Courtesy Rocky Mountain/Orthodontics.)

Fixed Appliances

The basic element of all fixed appliances is the orthodontic band. This is the metal band that is cemented around the crown of the tooth and to which are welded various types of brackets (Fig. 4-8). The bracket is the attachment by which the arch wire is affixed to the band. The bracket is precisely formed in many styles to fit the particular arch wire chosen for use.

The arch wire is the muscle of the appliance; it is a high-tension stainless alloy (occasionally precious metal) which, when expert hands connect it in various ways to the bracket, produces the pushing, pulling, rotating forces required in treatment while maintaining exact control over every individual tooth to which it is attached. The arch wire is affixed to the bracket in one of several ways; some brackets have lock pins, and others have locking caps that slide over the wire, but probably the most commonly used method is the ligature (usually referred to as "tie wire"). This is a fine steel wire that is hooked into slots in the bracket, embraces the arch wire, and is then twisted into a "pigtail" that is partially snipped off and the remainder tucked between arch wire and bracket. Assorted hooks may be welded to the bands to receive elastic bands. By using elastics of different sizes to interconnect the upper and lower arches, or different teeth within the same arch, a still wider range of movements is possible. These are the elastics that the patient forgets or refuses to wear and that pop out in your face when the patient laughs or sings.

It is primarily in the refinement of fixed appliances that the controversies previously referred to have arisen. the six major systems or methods now in use are (1) the round wire, (2) the twin wire, (3) the edgewise wire, (4) the universal appliance, (5) the labiolingual arch wire, and (6) the light wire.

Round Wire. The original concept of a multiband procedure was to connect the bands with a single round archwire that would supply desired forces. It moves teeth, but the crowns easier than roots.

Twin Wire. This involves the use of two light wires lying parallel in the same bracket rather than one heavier wire and achieves greater physiological tooth movement (Fig. 4-9).

Edgewise Wire (Fig. 4-10). One of Angle's early contributions was a "ribbon," or flattened wire, rectangular in cross section, that could be adjusted to achieve greater control of the apices of the teeth. he later discovered that by turning the ribbon "edgewise," using brackets machines so that the rectangular wire was inserted with its long dimension horizontal, he achieved better control over individual teeth.

Labiolingual Arch Wire (Fig. 4-11). Only molars are banded, two in each arch, with tubes instead of brackets welded to the bands. A single, heavy, round arch wire is fitted on the labial surface near the gingiva and attached by inserting the ends into the tubes. A similar arch wire is usually placed on the lingual surface. Movement is accomplished by means of vertical spurs and finger springs welded to the arch wire. Inevitably, there are many innovations and variations.

Light Wire. The light wire method is usually referred to as the "Begg light wire technique," or "differential force technique." It was developed in Australia by Dr. P. R. Begg in a neighborhood somewhat less than opulent and where high-tension wire, common in the United States, was difficult to find. He devised a method of using wire of small gauge,

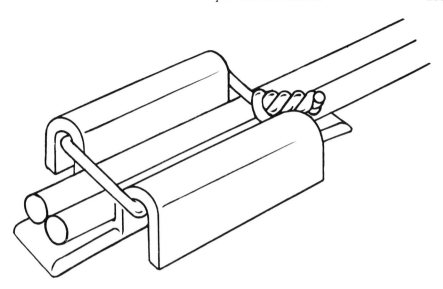

Figure 4-9. Twin wire technique. Note style of bracket, ligature wire, and "pigtail." (Courtesy Rocky Mountain/Orthodontics.)

which he bent into a series of coils, box loops, round loops, etc., and bracketed lightly to the bands in a manner that allowed the convolutions of the wire to press in various desired directions.

The Begg technique consists of three stages:

1. The crowns of all teeth to be moved are tipped simultaneously.
2. Spaces remaining after Stage 1 are closed.
3. Mesiodistal uprighting of teeth and labiolingual torquing of roots are accomplished by forces generated by torquing auxiliaries and spring-pins (Johnston, 1985).

The Begg technique requires true expertise in evolving its meanderings and has a surrealistic appearance, but with the development of improved wire it has been adopted by a large number of orthodontist.

Surgical Orthodontics

White and Proffit (1985) cite Public Health Service data that estimate 250,000 Americans need mandibular surgery for correction of Class III jaw relationships. Of these, an estimated 100,000 also need surgical maxillary advancement for correction of Class III malocclusions. An additional 750,000 are in need of both orthodontics and jaw surgery to

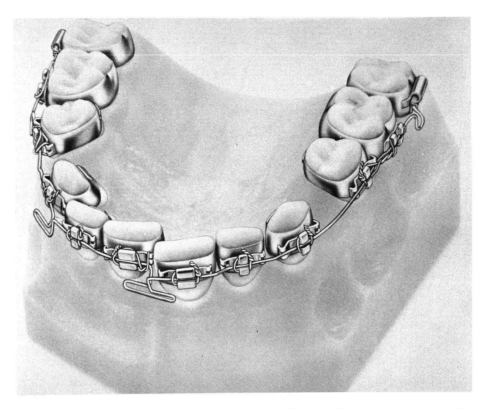

Figure 4-10. Upper arch banded with edgewise appliance. This is extraction case: first bicuspids have been removed. Note tube welded to bands on second molars; these receive labial arch of headgear. (Courtesy Rocky Mountain/Orthodontics.)

correct their Class II relationships. These writers discuss four developmental periods when surgery might be indicated:

1. During primary dentition: only for patients with congenital defects, such as cleft lip and palate.
2. During mixed dentition: surgery is generally contraindicated, due to unpredictability of growth patterns.
3. During early permanent dentition: Some surgery is indicated from *deficiency* problems. Most surgery, though, should be postponed until the fourth period.
4. During late permanent dentition, after growth is essentially completed: most surgery should be done here, and particularly for excessive growth correction.

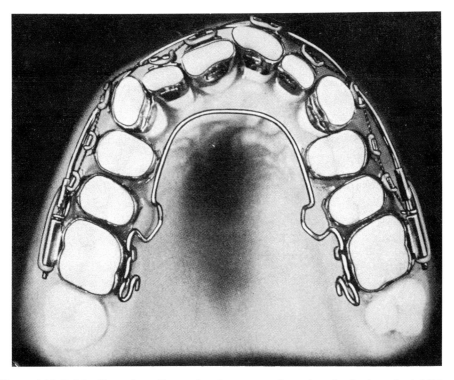

Figure 4-11. Labinolingual appliance showing many adaptations in single arch. Additional teeth are banded and other modifications incorporated. (Courtesy Rocky Mountain/ Orthodontics.)

Procedures

A surprising number of people phone an orthodontist, brush their teeth well, and keep their appointment with the expectation that they will walk out with a full-banded appliance in their mouth. Actually, it will be a few weeks to a few years before the bands make their appearance. It might be helpful to "walk through" a hypothetical case and examine what transpires between that first phone call and the removal of the last appliance some years later.

The first appointment in many cases is merely a matter of getting acquainted, while the orthodontist makes a gross evaluation of the problem, compiles some case history data, observes the general oral status and function of the patient, and, if myofunctional anomalies are found, refers the patient forthwith for correction of deglutition, a sucking habit, in-competent lips, or other obstacles in the orthodontic path. The dentist may then wish to delay further action for a few months, waiting for molars to complete eruption, for noxious habits to be eliminated, or for other reasons. In areas where there is an overabundance of ortho-dontists, the tendency is to band the patient first and ask questions later.

Records and Cephalometrics

The orthodontist usually does not want specific measurements of the patient until treatment is imminent. Accordingly, a later appointment is made for this purpose, generally called a record session or "taking records." At this time the dentist acquires the raw material for a detailed evaluation of the patient. Profile and full-face photographs are taken, impressions are made of both dental arches for construction of study models (record casts), and intraoral radiographs are taken of each seg-ment of the arches.

Often a second set of models is made, called working models: the teeth from this set can be cut off and remounted in a malleable plastic, making it possible for the dentist to manipulate them as he simulates the effect of the treatment he is planning. An oriented roentgenographic lateral headplate is made, and such other information as the dentist may desire is obtained.

The oriented x-ray film of the full head provides the dentist with invaluable information, but it is knowledge that must be teased out. The film is traced onto paper, and certain fixed points, or landmarks, are located and noted. By analyzing these points, by connecting some of them with others and measuring the angles of the lines thus formed, by noting relationships of teeth, jaws, and cranium, and by combining dimensional and angular criteria, the dentist is able to identify abnor-malities of both form and function, project growth and development, and thus diagnose and plan treatment. Later x-ray films can be superim-posed to check progress.

Computer Analysis

A diagnosis can be made only by a qualified clinician evaluating a live patient. Nevertheless, some of the tedium can be removed from the process described above. Some orthodontists now utilize a computerized cephalometric analysis program. The dentist submits a data sheet accompanied by both lateral and frontal x-rays; these data are examined, the films are analyzed by technicians, and over 50 points of measurement are located and translated into "computer talk." All of this material is fed into a computer, where it is compared to stored data from thousands of patients. The resulting print-out contains complete details of both diagnosis and treatment planning, including projected growth and development expectations both with and without treatment. Another set of x-rays is analyzed at the end of treatment to evaluate results. A number of other services are available in this program; the most important for the orofacial myologist is the consideration that all of this system is made available for research under proper conditions.

Temporomandibular Joint Analysis

As the orthodontist analyzes his records and begins to plan his treatment and establish goals for a particular patient, s/he finds almost invariably that certain compromises must be made. Some physical characteristics do not square with others; correcting one problem may preclude the repair of another—it may be absolutely essential to gain 4 millimeters of space whereas the only tooth can safely be extracted is 7 millimeters in width, etc.

While making these decisions the ultimate facial appearance of the patients must be kept in mind, some ideal which will be approached as nearly as conditions permit. Angle chose Apollo Belvedere as the epitome of facial symmetry, and attempted to build this profile into each patient. In the early days, esthetics was almost the only goal of orthodontic treatment, along with the effort to place the occlusal surface of each tooth in static contact with its natural antagonist. Given the fixed nature of the maxillary arch, this often resulted in a mandibular placement far from harmonious with other cranial structures, rendering serious insult to the temporomandibular joint (TMJ).

Orthodontists are studying the effects of their treatment on the requirements of the TMJ. It is apparent that serious health problems, of a seemingly unrelated nature, may readily supervene when even moderate stress proves intolerable to the temporomandibular joint over a period of years. The conscientious orthodontist takes this into account as he studies and plans. An orthodontist who restricts practice to problems of the TMJ is called a gnathologist.

Conference Appointment

Some time after records are taken, often two or three weeks, the orthodontist has made a full analysis and is ready to discuss his plan of treatment with the patient or the patient's parents. A discussion period is arranged, during which the dentist explains what will be done, how it will be accomplished, how long it is likely to take, what it will cost, and manner of payment; usually a contract is signed containing many of these provisions. The dentist will request that any cavities be repaired before treatment can begin, and outline the cooperation expected of the patient and parents. This may be the last chance that parents have to ask questions directly of the dentist, but unfortunately they seldom know which questions to ask; future information will be exchanged via the child, who often forgets or gets it wrong.

Extractions

During the above conference, the dentist will also reveal decisions as to extractions. S/he will have measured the total length of the bone in each arch, and added up the combined width of the teeth within each arch. When the arch perimeter is inadequate in which to place each tooth in a healthy position, when permanent teeth are ankylosed, or when one or two other conditions prevail, the patient may be referred to an oral surgeon for extractions, and an "extraction letter" sent specifying exactly which teeth are to be lost, usually the first or second bicuspids.

It is not sound orthodontics to extract only one or two teeth as a means of gaining space; if one bicuspid goes, four should go, one from each quadrant, since the symmetry of the arch must be maintained and the teeth in the two arches must be capable of intercuspation in a reasonably normal manner at the conclusion of treatment. There are exceptions though, such as when inter- and intraarch asymmetries need correcting. Nonextraction treatment is invariably to be preferred, when it is possible; the sacrifice of permanent teeth is not a step to be taken lightly, although a few dentists treat almost every patient as an extraction case, robbing the tongue of essential space and sometimes creating a "dish face," in which the midportion of the face has a concave appearance.

In any event, extraction sites must heal before an appliance is installed. With careful timing, it is often possible to initiate myofunctional therapy at this juncture, as long as it can be completed before the appliance is scheduled.

Expansion Devices

Some patients, particularly those displaying certain types of oral muscle dysfunction, exhibit what has been called a "collapsed upper arch," a palatal vault where the lateral segments of the basal bone itself have developed so little lateral dimension as to be incapable of containing the tongue or of supporting the teeth in any sort of normal alignment. In these cases, some orthodontists turn to a procedure called "splitting the palate," or R.P.E. (rapid palatal expansion). Devices for expanding the arch vary somewhat, from a pair of turnbuckles embedded between two blocks of acrylic, to the somewhat lighter wire arrangement shown in Figures 4-12 and 4-13. The patient is supplied with a key with which to rotate the expansion screw a prescribed number of turns each day, breaking open the midline suture of the palate and thus creating a greater horizontal dimension. This can usually be achieved in only two to three weeks, after which a retaining appliance is installed, or the expansion device simply left in place, until the suture area has recalcified as it is filled in by the deposition of new bone.

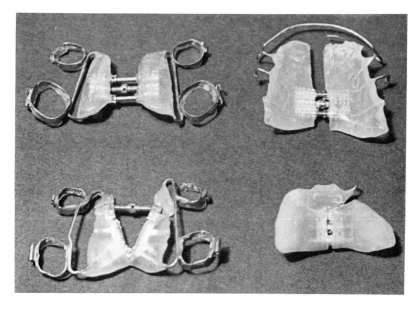

Figure 4-12. Palatal expansion devices. Each of those pictured incorporates an acrylic body, but others are made entirely of metal. A, Expansion screw has been turned to its fullest extent. B, Labial wire rather than bands. C, Hinged at posterior extremity, this expands anterior palate. D, Cut from its original bands, this has second expansion screw designed to move anterior segment forward, while expanding buccal segments laterally.

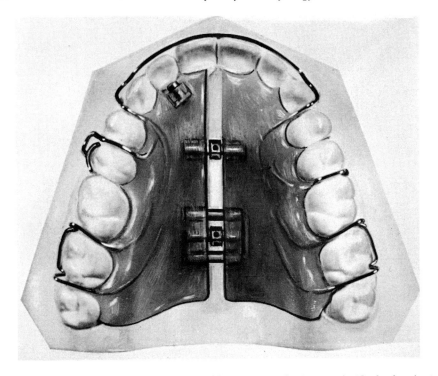

Figure 4-13. Expansion device in place. (Courtesy Rocky Mountain/Orthodontics.)

Again, this is not a procedure that can be used indiscriminately, for there are at least two conditions in which it is contraindicated. One occurs when the upper molars still occlude in a normal lateral relationship with the lowers. Unless the molars are in bilateral crossbite, expansion may set the maxillary teeth so far buccally as to make any normal relationship with mandibular molars impossible; it would be necessary to tip the lower teeth at a most unphysiological angle to accomplish any kind of interdigitation.

The other inadvisable condition results from the abnormal muscle function itself that so frequently accompanies these cases. We will stress in later chapters the fact that any untreated malocclusion is in a state of dynamic muscular balance. When maxillary teeth are suddenly thrust against the musculature of the buccal wall, with no provision for compensatory alteration of the environmental forces, there is risk of relapse or even greater damage to the teeth.

Should the patient be scheduled for R.P.E., the oral myologist might safely carry out some preliminary lip and cheek procedures, but should never attempt to complete therapy for abnormal deglutition until the expansion device is removed. When therapy is extended prematurely, the realization comes rather quickly that the appliance itself forces the tongue into an abnormally low posture. This prevents any semblance of normal function in swallowing, and thus assures a return to tongue thrusting, even in the relatively brief time that the appliance is in situ.

It has seemed more logical to wait until the expansion device is replaced by a retainer or full bands. The retainer can be removed for practice sessions, interferes little with routine function, and once the tongue begins to exert palatal pressure it makes an excellent substitute for the retainer.

Separation of Teeth

The patient might reasonably expect, after submitting to all of the foregoing preliminaries, that s/he is finally ready to acquire bands, but it is not to be. S/he has yet to make the acquaintance of one of the less pleasant aspects of treatment, the separator wire.

Even in severe malocclusion, most of the teeth are pressed tightly together at the contact point. Still, the orthodontist must have the thickness of two bands and two layers of cement, one over each contact point, between each tooth and its neighbor on either side in order to place the bands. The answer lies in separation of the teeth.

There are several methods by which to accomplish this. Probably the most common is to place small loops of brass or stainless steel wire around each contact point, then twist the ends together, drawing the loop tightly around the contact, and tucking the "pigtail" into the gingival embrasure. Another common procedure is to use a commercially available product, a tiny molded, bone-shaped piece of elastic material; when grasped at each end and stretched, it can be drawn between the contacts and released, wedging them apart as it returns to original thickness. Other dentists employ elastic thread or very small elastic bands for this purpose. The teeth are usually separated within two or three days, and the separator devices begin to fall out, frequently to be ingested. They stop hurting after the first day.

Banding the Teeth

The separator session is frequently combined with a fitting session, during which bands are selected for each tooth and grossly fitted; they are adapted precisely at this, the banding appointment, after which the brackets are welded in place and the band cemented to the tooth.

Great variation occurs at this point. Some dentists prefer to band all teeth in one marathon session; others band one arch and wait a week or more for the other, while some install one segment at a time over several weeks or months: upper molars, lower molars, upper anteriors, lower anteriors, etc. Some place and activate the arch wire as soon as bands are in place, while others attach it only loosely at this time. Still others wait a week or so before installing the arch wire.

A few factors in addition to personal preference enter in, but most orthodontists follow a customary routine that the oral myoclinician does well to learn: when active therapy has been completed before banding, it is advisable to recheck the status within two weeks after the sudden dramatic change in the orofacial environment occasioned by a full-banded appliance.

There is an inclination to live life without dental occlusion for a few days after banding, in the fear that biting will increase the discomfort. In fact, those who choose hamburgers over soup for their first postappliance meal seem to fare much better; the sooner the jaws are closed with some force, the quicker the bands are forgotten. Danger lies in the proclivity of some patients to place the tongue interdentally as a cushion, a precaution against unexpected closure; newly acquired swallowing and resting patterns can deteriorate rather quickly under this influence.

Once banded, the patient begins the routine of returning every two to four weeks for adjustment of the arch wire as the teeth are slowly moved into alignment. (The mechanics of how this is accomplished is afforded a special section below.) For now, we may note that by bending the arch wire into various configurations, and combining its force with elastics, headgear, or other supplemental pressures, the desired changes are effected.

Innovations

One relatively recent development has been the appearance of the "arch wire with a memory." This is an alloy called Nitinol made of nickel and titanium; it is drawn into a fine wire, .019 inch in diameter, which is quite flexible and easy to manipulate. It is then formed into the desired curve, heated briefly to 800° C., then quenched in cold water.

It can thereafter be bent to conform to the present shape of the patient's arch and ligatured in place. Body heat is sufficient to trigger the efforts of the wire to return to the shape built into its memory, exerting desirable force not found in conventional alloys or stainless steel arch wire.

Claims for Nitinol include the ability to reduce treatment time by 50 percent for the patient while increasing comfort, since it is not necessary to tighten it so much or to replace it every month or two. The orthodontist saves valuable chair time otherwise spent in arch wire changes.

Nitinol was developed by researchers in the space program, where it has several applications. It was brought into orthodontics by Dr. George F. Andreasen, Chairman of the Department of Orthodontics at the University of Iowa. Considering the long history of the Andresen appliance, Andreasen may prove a trip-word for the unwary.

The cosmetic affliction that some patients feel as a result of wearing bands, and the revulsion that comes with time when they are repetitiously labelled "metal mouth," "railroad track," "brace face," or "tin grin," has spurred some orthodontists to develop less unsightly appliances (Fig. 4-14, A). One or two will be described in case the clinician is suddenly faced with something unfamiliar.

Direct bonding is a concept that reappears at intervals and is now in a resurgent phase because of improved techniques and materials. This involves etching the tooth by washing the surface with an acid solution, thereby clearing the microscopic crevices in the enamel of all foreign matter, then bonding the bracket directly to the tooth without benefit of any type of band (Fig. 4-14, B). This results in some loss of control in moving roots, but can be effective when both patient and dentist are careful. The brackets pop off when the patient chews hard or coarse food, and sometimes when s/he does not.

The effect can be made more "invisible" by using clear plastic brackets bonded to the tooth with a clear adhesive that leaves only the arch wire a metal feature. This demands some additional compromise, since the arch wire can distort and cut the bracket, and excessive force can fracture brackets. Nevertheless, there are a number of advantages to direct bonding, whether with metal or plastic brackets; tooth separation is unnecessary, as is closing spaces after band removal, the uncomfortable force-fitting of tight bands, etc.

Indirect bonding has been tried using several methods. The effort grew out of some dissatisfaction with direct bonding, in which human limitations often influence the method. That is, when the bracket is placed against the tooth in direct bonding, it may be turned or moved just slightly from its optimal position; this throws awry the action of the arch wire. Accordingly, dentists have experimented with placing the brackets in ideal alignment on the working models but only lightly

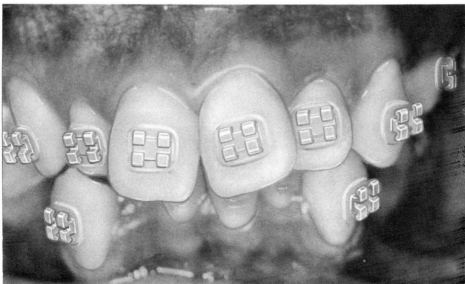

Figure 4-14. Direct bonding. A, Patient wearing "almost invisible" appliance. B, Plastic brackets bonded directly to teeth. (Courtesy Rocky Mountain/Orthodontics.)

attached so that errors of placement can be corrected. The brackets can then be embedded superficially in a matrix of some sort and transferred in toto to the real teeth of the patient. Given an orthodontist who is part magician, it works.

Straight wire appliances are beginning to appear in some quarters. Using conventional methods, the orthodontist spends an appreciable amount of his/her time in bending arch wire in and out, up and down, in order to achieve the desired movement of the teeth. The straight-wire concept is that most of this bending can be eliminated by using a system of brackets which has the necessary angulation "built in." That is, these brackets are made in varying thicknesses to stand out from the band to whatever degree required by the individual tooth, and are precisely tipped to accomplish torquing, angulation, and other details of tooth positioning. Most of the shaping and forming of the arch wire is then unnecessary, a relatively straight wire can be ligatured in place, and the treatment is built into the bracket. These brackets are also being produced for use with direct bonding techniques.

Tooth Movement

Types of Movement

There are five general types of tooth movement used in treatment: tipping movement, bodily movement, rotation, extrusion, and intrusion. It should be borne in mind that the orthodontist is not merely uprighting teeth into a previously prepared slot, as if moving a gearshift lever. He/she is literally moving the entire tooth, complete with alveolus, through bone. This maneuver is accomplished through the physiological process of resorption and deposition, shown in Figure 4-15. Bone is vital, living tissue undergoing constant revision and is more responsive to force than is often appreciated. Pressure against the bone causes resorption of existing bone until the pressure is eased; tension or pulling on the surface results in the deposition of new bone until a balance of forces is restored.

Orthodontists usually refer to this process as "physiological tooth movement." A different appreciation of their work may be gained through the realization that to benefit the patient, the orthodontist is actually injuring the tissue, inflicting a deliberate but controlled pathology. The periodontal membrane governs the consequences. Pressure against the

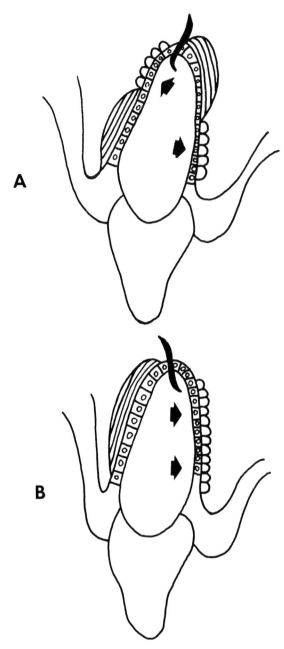

Figure 4-15. Types of tooth movement. Results of pressure applied lingually to crown of central incisor (periodontal membrane greatly enlarged). A, Tipping movement: periodontal membrane is crushed at alveolar crest lingually and at apex labially; fibers are stretched at opposite diagonal. Scalloped lines show bone resorption, and concentric lines indicate bone deposition. B, Bodily movement: entire membrane is compressed lingually and stretched on labial side.

crown of the tooth is transmitted through the root, reducing circulation in the periodontal membrane in areas where the membrane is compressed while straining its attachments in other places. The reaction of tissue to such insult effects tooth movement.

Tipping movements are relatively fast even with moderate pressures, for the periodontal membrane recovers quickly; bodily movement, compressing a much greater portion of the membrane, requires a longer recovery time. It is thus common for the patient wearing an appliance effecting bodily movement to see the orthodontist only at three-week intervals for adjustment.

Total time during which the appliance is worn may vary from a few months up to two years or slightly longer, but long treatment periods of three to eight years often indicate an untreated tongue thrust or some other etiological agent.

Mesial Drift

Mesial drift is nature's orthodontic procedure, not the dentist's, but it is a condition that the dentist must control at times. It is the tendency of the teeth to shift toward the midline of the dental arch until contact is established with another tooth. The forces at work in the normal mouth militate to this end, so that if a tooth is extracted, the teeth distal to the resulting space "drift" mesially through the bone by the natural occurrence of resorption and deposition, as described above. As neighboring teeth rub together in function, the proximal surfaces wear away and flatten; mesial drift keeps them in contact, nevertheless. When a deciduous tooth is lost prematurely, we have noted that a "space maintainer" is frequently employed to combat mesial drift and to hold a place in line for the future arrival of the permanent tooth destined for that space. Mesial drift can be a beneficial force when permanent teeth must be extracted. Abnormal tongue pressures can prevent mesial drift in extraction cases, causing an undesirable spacing of the teeth.

Force

The optimum force that should be applied in order to move teeth has not yet been established; it will be a difficult formula to devise, for many unknown factors remain. One proposal has been 20 to 26 grams (less than 1 ounce) per square centimeter of root surface, the capillary pulse pressure of the body through which teeth erupt and drift naturally. In actual practice a greater force is often used, but it must be kept under rigid control to avoid such unpleasant sequelae as root resorption, sheared alveolar crests, and poor gingival health.

Age

The age factor is of vital concern to the dentist, for he must utilize growth processes to achieve some of the necessary changes. The growth spurt that usually occurs when a child enters puberty is often prime time to the orthodontist. The apices are fully formed and closed, so that major tooth movements are possible without danger of root impairment. Prior periods of rapid body growth are also precious, but the puberty cycle is the "last chance" timing for some aspects of Class II and Class III treatment. That is, this cycle essentially completes the growth of the basal bone forming the jaws. Thereafter, the orthodontist is not able to take advantage of growth by restraining development in one arch while the bone of the opposing arch "catches up" as it continues normal growth and development. Such restraint or encouragement of growth is the purpose behind some uses of the headgear appliances discussed earlier.

By no means does this indicate that there is no help for the adult patient; remarkable changes have been effected in patients over 40 years of age when orthodontic treatment followed correction of the swallowing pattern. Individual teeth have been moved for 70-year-old patients in order to place a prosthesis.

Anchorage

"For every action, there is an equal and opposite reaction." This law of physics will concern the therapist more directly in matters of deglutition, but it is basic to the dentist in tooth movement. To apply force to a tooth, there must be a point of resistance to that force, a "skyhook" by which to lift. In the past, it was usual to wait until the permanent molars were present in the mouth and to use their more securely affixed root structure to effect movement of anterior teeth. However, some seven or eight types of anchorage have now been devised, including the extraoral resistance inherent in the headgear in which the neck or occiput supplies anchorage.

Retention

Joy is unrestrained on the day bands are scheduled to be stripped from the teeth. Lips that have been drawn together, concealing the appliance, are now rehearsing broad smiles, to exhibit beautiful unencumbered teeth. Still, one important phase remains to be completed: the retention period.

Once the bands are stripped off, the teeth are usually buffed and polished, and some occlusal equilibration is often done to encourage healthful intercuspation and make other fine adjustment. Typically, at this same appointment, "finish models" are taken, sometimes for later comparison with the original study models but also as a guide for constructing retention appliances. The patient will return two or more days later to receive retainers, as it requires a few hours to fabricate them.

Extensive changes have been wrought in the tissues of the mouth as a result of orthodontic pressures. We referred earlier to orthodontic treat-ment as "controlled pathology." Just as the orthopedic surgeon must splint and brace his/her handiwork until bones have knit and tissues healed, so too the orthodontist must immobilize the reorganized mouth until it is able to withstand environmental forces unassisted. Additionally, spaces between the teeth required for the bands must now be closed.

We have pointed out the great diversity of orthodontic approaches and appliances. The dentist understands quite well the physiology of tooth straightening, and so is able to arrive at the desired result along many different routes. However, the actual body of scientifically proven data concerning retention is discouragingly small, with the result that there is remarkable uniformity in retention procedures: a cuspid-to-cuspid fixed bar behind lower incisors and a Hawley-type removable on the upper arch. It has been done this way for many years, and few dentists challenge the system.

The lower appliance is made by rebanding the long-rooted lower cuspids and connecting the two bands with a light bar of stainless steel adapted closely to the lingual surface, thus preventing any sort of distal movement of the lower incisors. If an anterior crossbite has been corrected, the upper teeth themselves act as the retainer.

The upper retainer varies somewhat in design, depending upon whether or not there have been extractions. With acrylic fitted precisely to the lingual surfaces and a static wire around the labial surface, the upper teeth are effectively locked in place.

Occasionally an orthodontist will remove the bands earlier than s/he would otherwise do, and use a positioner as both a "finisher" and a retainer. The teeth from the finish models are cut off and mounted in wax as described earlier, so that the impression of each tooth within the positioner reflects ideal position; the patient bites into this guidance

system four or five or more hours per day while awake and continues
through the night. Positioners are frequently less acceptable to the patient,
and in most cases must still be followed by the conventional retainers
described above.

The length of time during which retainers are worn depends on two
factors in addition to the preference of the dentist. One consideration is
the type of tooth movement that has been carried out; teeth that have
been rotated require much more time than those that have been shifted
or tipped. The second factor is the age of the patient. When treatment
has been concluded during body growth, much of the necessary reorganization of tissue has occurred spontaneously as a part of the growth
process, reducing the time required for retention. Adult patients must be
retained over a longer period in order to be secure.

The duration of the retention period may thus vary from a few months
to several years. Patients have been seen who have been in retention for
20 years or more; we might add that the authors have not seen such a case
who displayed a healthy mouth or even reasonably normal occlusion.
Long retention is demanded only when teeth are placed in a posture
which is inharmonious with their environment. The antagonistic stress
generated even by normal surrounding tissues can result in thickening
of the periodontal membrane and other ills leading to dental breakdown.

Whatever the continuance of retainers, when the last appliance is
removed the patient's musculature resumes command of tooth posture.
Given normal function, a beautiful, healthy, permanent result emerges;
it has been the conviction of some orthodontists that retention devices
should not even be required in the thoroughly treated mouth.

Orthodontic Relapse

In contrast, an untreated tongue thrust makes its presence known
rather quickly at this juncture. No amount of retention is adequate when
the dentition is subjected to abnormal environmental forces. Some dentists maintain quite a vocabulary of euphemisms for use in these cases:
"settling," "post-treatment adjustment," "God's will," and "preeminence
of the morphologic pattern"—any of these sounds less offensive than the
term *orthodontic relapse.* Mason (1988) warns readers that too often the
tongue is unfairly blamed for orthodontic relapse. He offers three basic
reason for the necessity of the retention phase of orthodontic treatment:

1. Time is required for reorganization of the gingival and periodontal tissues that were moved during active orthodontic treatment.
2. Additional growth following orthodontics may significantly alter teeth and jaw positions.
3. The teeth may be relocated by orthodontics in an unstable position so that the soft tissue pressures constantly encourage a relapse tendency.

We should make one final point concerning retainers. These appliances are designed to resist force only on a horizontal plane, but exert no vertical pressure and are thus incapable of counteracting the effect of a tongue placed interdentally. Open bite cases may begin to reopen with dramatic suddenness even with retainers in place, once relieved of the interarch elastics used to close the bite.

REFERENCES

Graber, T. M., and Neumann, B. (1984). *Removable orthodontic appliances* (2nd ed.). Philadelphia: W. B. Saunders.

Johnston, L. E., Jr. (1985). *New vistas in orthodontics.* Philadelphia: Lea and Febiger.

Mason, R. (1988, March). Orthodontic perspectives on myofunctional therapy. *Intrn. J. of Orfacial Myology, 14,* 1. (Special Edition)

White, R. P., Jr., and Proffit, W. R. (1985). Surgical orthodontics: A current perspective. In L. E. Johnston, Jr. (Ed.), *New vistas in orthodontics* (260). Philadelphia: Lea and Febiger.

CHAPTER FIVE

ANATOMY FOR THE OROFACIAL MYOLOGIST

Anatomy is an area of absorbing interest to the nonanatomist only in the context of gender. It is a subject like basic mathematics (and unlike spelling) in which it is hard to be original or creative—basic anatomy changes little. We strain our mind almost to the breaking point in attempting to understand the workings of the human mind. Mostly, anatomy is memorization.

There is no scarcity of textbooks on either anatomy or physiology, some of which run to over 1200 pages and many of which are specialized in nature and thus able to eliminate extraneous material and focus on a certain portion or system of the body with some depth. For those who have not established friendly relationships with one or more of these books, especially those exploring the head, neck, and torso, we heartily recommend making their acquaintance. Yet we know of none that relates specifically to the act of deglutition or the vagaries and associated concerns thereof. Hence we will devote a handful of pages to this pursuit.

As with several of the preceding chapters, each of which is intended for only those readers whose needs include the specific category of information given, this will be a sharply circumscribed presentation of anatomy and physiology. We will look backward through the large end of the telescope at only those structures that have pertinence for the orofacial myologist. However, we will try to explain *why* and *how* these elements relate to oral function.

There is a threat of gross misunderstanding when we delimit so drastically the extent of our inspection. It is essential that there be an ever-present awareness of the *oneness* of the miraculous human body, every fiber affecting and being affected by every other cell as well as the sum of the parts. The function of an oral muscle may be influenced by tension in a remote or seemingly unrelated muscle, by body posture or temperature, by emotion or state of mind, by diet, and by many other factors. Such considerations will be laid aside for the moment, along with

123

the skin and flesh that we must strip away, in order to view the tissues we seek and to offer a brief and partial explanation of their purpose.

Terminology

As a general rule, the language of anatomy is specific, requiring some agreement among discussants for full understanding. Variations do occur; Latin names are often Anglicized (as will be the case here in most instances) but we will define a few terms essential to an orderly perusal of this material. In addition, we will conclude this chapter with a brief glossary of anatomic terminology. In order to avoid duplication, we will eliminate terms recorded in the dental glossary at the end of Chapter 3. Should we unintentionally use a term not shown in either glossary, desperation may drive the reader to one of the formal texts listed in the references.

Viewing the erect body face-to-face, we should establish our directions, for some minor reorientation is necessary for readers with a dental background. The dentist's "mesial" becomes the anatomist's "medial" and no longer accommodates the concept of traversing a curved dental arch, but instead implies a direction squarely toward the vertical midline. Its opposite becomes "lateral," denoting direction away from the vertical midline. The other terms that alter in meaning as they commute between dentistry and anatomy are "proximal" and "distal." Proximal, as used in this chapter, indicates direction toward a point of attachment, while distal denotes the opposite, or away from the point of attachment; these words are used primarily in describing features of the limbs, to indicate relative distance from the attached end.

Four other words should be clearly understood; the meaning of these is the same in both fields:

ventral (anterior) Toward the front.
dorsal (posterior) Toward the back.
cephalad (superior) Toward the head.
Caudad (inferior) Toward the tail or lower end.

In order to examine the interior of the body and some of its organs, it is helpful to divide the structure into sections, removing one section for easy viewing. The direction of these cuts, or *anatomic planes,* must be clear.

Sagittal Plane. A division vertical to the ground and front-to-back, yielding right and left parts or sections. When the division is made to coincide with the vertical midline, it is called the *median* or *midsagittal* plane.

Frontal or **Coronal Plane.** A lateral or side-to-side division, vertical to the ground, separating the front section from the back.

Horizontal or **Transverse Plane.** Division across the structure, parallel to the ground, forming upper and lower sections.

Divisions of Anatomy

The study of the structure and function of the human body is made more meaningful, efficient, and economical in terms of time and effort by dividing the total into cohesive units of special interest. Were we concerned solely with muscle function, we might confine our attention to *myology,* that segment of anatomy devoted to the study of muscles. Such an exclusive perusal would have little meaning, however, without some knowledge of *osteology,* the study of bones; most muscles derive their name from points of attachment to the skeleton or other body structures, and their function can be understood only in relation to their effect on the tissues thus connected. It is also helpful to have some concept of *neurology,* the study of nerves within the nervous system that control the muscles, and *angiology,* the study of the vascular system supplying blood to the muscles. The applicable details of both neurology and angiology are readily available from many other sources, and therefore we will not duplicate them herein.

Among the many other divisions into which anatomy is segmented are: (1) *cytology,* the study of cells; (2) *histology,* the study of cell groups as they form the component parts of the body; (3) *embryology,* the study of the growth of cells from the fertilized ovum to full differentiation; and (4) *syndesmology,* the study of joints or articulations. Other specialized areas are concerned with the *visceral system,* the *respiratory system,* the *digestive system,* the *urogenital system,* etc.

Some Types of Tissue

Groups of cells joined for a common purpose and functioning in harmony form a tissue. Tissues come in many types, sizes, and compositions. In fact, we may view the human body as a fantastically intricate mass of differentiated tissues, such as cartilage, bone, blood, and connective tissue (all of the preceding are sometimes referred to as connective tissue) held in place and set in motion by muscular tissue, which in turn is activated under the governance of neural tissue.

While confining our attention to the area of the head and neck, we will examine these tissues in order. We do so with the understanding that body tissue may be classified in several other ways and into many other categories. And we will save the inspection of muscle tissue until last, since a fuller comprehension and a broader view of their structure and function will then be possible.

We will not go into any great detail at this point concerning *connective tissue.* Those tissues that have an application to this field will be pointed out in the context of muscle function. For now, we may note that the connective tissues that we encounter will consist primarily of four types: loose, dense, regular, and special. All are collagenous, with abundant interlacing processes, and may pervade, support, and bind together other tissues and organs and form ligaments, tendons, and aponeuroses. *Loose* connective tissue is strong, inelastic, and white. *Dense* connective tissue is thicker, more compact, and more elastic. *Regular* connective tissue is composed of parallel bundles of fibers that have great resistance to stretching, and accordingly is the type found in tendons and ligaments. Connective tissue with *special properties* varies rather widely, its composition depending in part on its location and function: examples are mucous membranes, adipose (fat) tissue, the connective tissue of the periodontal membrane, etc., each differing in some respect from the others.

Cartilage

Cartilage is a dense mass with a very uncomplicated structure. It contains no blood vessels or nerves, receiving its nutritive elements from surrounding tissue. It is the precursor of bone in the embryo; the fetal skeleton consists primarily of "temporary" cartilage. Ossification occurs on a varying timetable, depending on the location and precise consistency of the cartilage. Much is complete at birth although the fontanels or "soft spots" of the neonate skill will require some further time; the transition to bone happens in "permanent" cartilage only with advanced age, or never.

Cartilage may be classified according to its internal structure into three types: hyaline, elastic (also called yellow fibrocartilage), and white fibrous.

Hyaline. Encompassing several subtypes, hyaline cartilage is the most abundant type. It has a shiny bluish-white appearance and has considerable elasticity despite being quite firm in texture. Hyaline cartilage encrusts the ends of bones that are encased in joints, its smooth surface providing ease of movement. Bars of hyaline cartilage connect the anterior ends of the ribs to the sternum or to each other, lending enhanced elasticity to the walls of the thorax. It also combines elasticity with stability in vital airway passages in the nose and trachea.

Elastic. For our purposes, we need only note that elastic cartilage, more elastic and flexible than hyaline cartilage and with a faint yellow tinge, makes up the epiglottis, much of the larynx, the eustachian tube, and the external auditory canal.

Fibrous. The color of fibrous cartilage is white, since the cartilaginous tissue is mixed with varying amounts of white fibrous tissue. The latter ingredient adds toughness to the elasticity of cartilage, making its appearance appropriate in joints subject to frequent or extended movement, and to joints that are exposed to sudden or violent shock. Therefore, it is hardly surprising to find this type of cartilage in the temporomandibular joint, the most frequently and stressfully used of all joints; even this tough, resilient tissue, in this location called the *articular meniscus,* is not always equal to the strain and abuse to which it may be exposed.

Fibrous cartilage, also found in wrist and knee joints, forms discs between the bodies of the vertebrae (where they also rupture on occasion), and forms a thin coating over the surface of the grooves through which the tendons of certain muscles glide.

Synovial Membrane. We cannot conclude this section without some general description of the structure of joints and the critical synovial membrane. When joints are immovable, as those between bones of the skull, the margins of the bones are separated only by a thin membrane, called the *sutural ligament.* When slight movement is involved, as in the joints between vertebral bodies, we have noted above that the osseous surfaces are joined by fibrous cartilage. However, freely movable joints present quite a different structure. In order to achieve a full range of motion, the surfaces of the bones are completely separated, this division sometimes being further increased by a rim of fibrous cartilage surrounding and thus deepening the articular cavities. That portion of the bone forming the joint is covered by cartilage, as previously described, the joint is strengthened by ligaments which attach to the articulating bones, and the joint enveloped by a capsule of fibrous tissue. The interior of the fibrous capsule is lined with *synovial membrane,* which secretes a lubricating fluid, facilitating the action of the joint.

Osteology

The adult skeleton consists of over 200 named bones; we must be concerned with only about one-eighth of these, but it is essential that we lift our eyes above an overfascination with mere maxillary and mandibular bone.

Bones may be classified according to their shape into four groups: *long*, found in the arms and legs; *short*, located mostly in wrists and ankles; and the two with which we will deal, *flat* and *irregular*. Flat bones exist where there is need for extensive protection, as in the upper skull. Irregular bones have peculiar and often individual shapes, as in the sphenoid bone of the skull.

Bone is distinguished from cartilage by having a very dense intercellular matrix containing mineral salts, mainly calcium compounds. Depending on its composition, bone is classed as either dense (compact) or spongy (cancellous). The outer portion of each bone is composed of a layer of dense bone, which has the appearance of a continuous hard mass, with a central portion in most cases containing a space called the medullary or marrow cavity. Spongy bone makes up all of a few bones and a portion of most bones, and consists of intercrossing and connecting osseous bars of various thicknesses and shapes; while appearing quite porous and flimsy, the arrangement of these bars provides the bones with maximum rigidity and resistance to stress or changes in shape, adding strength without adding weight. Regardless of its composition or structure, all bone is plastic and responsive to pressure.

The outer surface of each bone, except for articular surfaces, is covered by a closely attached layer of dense fibrous tissue called the *periosteum;* the periosteum is richly supplied with blood vessels and nerves, and plays an important role in providing blood and sensory innervation to bone.

A number of bones with which we are concerned are characterized by irregularities, projections, and depressions that have specific descriptive terms, some of which are defined in the accompanying box in order to clarify our discussion. We will confine our attention to the bones of the skull plus the hyoid bone, with a fleeting glance at an upper vertebra or two. Even some in this limited purview will be merely listed, despite their importance in other contexts.

The skull is composed of 21 bones joined closely together and one, the mandible, that is freely movable. This does not include the three tiny bones of the ossicular chain in each middle ear cavity.

It would be helpful at this point if we could supply each reader with an actual skull, for many features are easily determined by examining the skull as a whole; some are seen in Figure 5-1, which also identifies the cranial bones. The two large sockets housing the eyeballs are the orbits. Between and slightly below the orbits is the pear-shaped nasal opening, divided vertically by the nasal septum. The nasal bone itself appears quite short, without its fleshly extension of cartilage; but within the nasal opening on each side can be found three scroll-like bony processes, the nasal conchae or turbinates. The zygomatic arch, or cheek bone, flares from the side of the skull and circles around beneath the orbit. Behind the orbit and above the zygomatic arch, the external surface of the sphenoid bone swoops in to form the temple, or temporal fossa, with only minor participation of the temporal bone itself. The mastoid process is the bulky projection behind and below the external auditory canal. Viewed from below, the foramen magnum is the large opening through which the spinal cord emerges. The hard palate forms the floor of the nose, while the floor of the brain case can be seen from above to consist of three large uneven depressions called the anterior, middle, and posterior cranial fossae.

TERMS PERTAINING TO BONE

process. An arm or projection; any marked bony prominence
tubercle. A small rounded process
tuberosity. A large rounded process
condyle. A rounded, knuckle-like process
spine. A sharp, slender process
crest. A narrow ridge of bone
ala. A wing; a wing-like structure or appendage
head. A part supported on a constricted portion, or neck
fossa. A depression in or upon a bone
sulcus or **groove.** A furrow
fissure. A narrow slit
foramen. A hole or orifice through which blood vessels, nerves, and ligaments pass
meatus or **canal.** A long tube-like passageway
antrum or **sinus.** A nearly closed cavity or chamber in a bone, or one having a relatively narrow opening

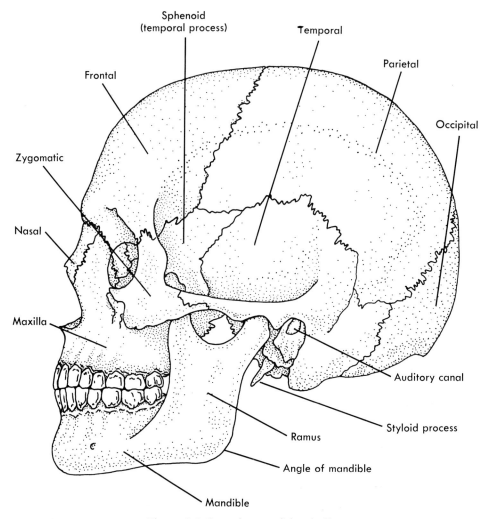

Figure 5-1. Some bones of the skull.

Cranial Bones

Eight bones are incorporated into the cranium: two parietal, two temporal, and one each frontal, occipital, sphenoid, and ethmoid. These blend into the fourteen bones of the facial skeleton, with little to mark the distinction. Since our interest in bone is primarily in its function of providing attachment for muscle, we will not discuss certain of the cranial bones.

Frontal Bone. An exception to the preceding statement is the frontal bone, the saucer-shaped bone forming the forehead and upper part of the orbit; it also forms part of the septum between the brain case and the nasal cavity. While it does serve as attachment for a few muscles, it also houses the frontal sinuses and nasal cavity. With our demanded concern for nasal breathing patterns, it behooves us to note that the frontal sinuses play an important part by expanding the area available to the nose in warming, humidifying, and filtering the inhaled breath.

The frontal sinuses are two large members of the group connecting directly with the nasal passage, a group known collectively as the paranasal air sinuses. The frontal sinuses are located superior to the orbits, vary greatly in size and shape, and empty into the middle meatus of the nose. Their contribution will be detailed later in this chapter, after we encounter some of the other sinuses.

Temporal Bones. On either side of the cranium are the two temporal bones forming part of the lateral wall of the base of the skull. The temporal bone is usually divided for study into five parts; the *squamous, petrous, mastoid, tympanic,* and *styloid* portions.

The squamous is the large, fan-shaped portion that forms a part of the temporal fossa, and from which arises the temporalis muscle. Projecting from the lower part of the squamous region is the long arch of the *zygomatic process,* the inferior border of which provides origin for the masseter muscle. The zygomatic process articulates with the temporal process of the zygomatic bone.

The petrous portion is shaped like a pyramid and is wedged in at the base of the skull between the sphenoid and occipital bones. The middle and inner ear are contained in this portion, along with other features of the auditory system. On its inferior surface is the origin of the levator veli palatine muscle, part of the bony portion of the cartilaginous eustachian tube, as well as the root of the styloid process. Just anterior to the styloid, where the petrous and squamous portions join, is the round, smooth *mandibular fossa* for the lodgment of the head of the mandibular condyle.

The mastoid portion lies inferior to the squamous, and is the most posterior part. It forms the bulky prominence that can be felt behind the ear. Beneath this portion is the *mastoid process,* a large cone-shaped projection which provides attachment for the digastric and sternocleidomastoid muscles, among others.

The tympanic portion forms most of the bony portion of the external auditory meatus (canal) and provides attachment for the cartilaginous segment of the canal. It also contains the tympanic sulcus, a groove which receives the tympanic membrane (eardrum).

The *styloid process* is a long, slender, bony spike projecting downward and forward from the temporal bone. This is the point of attachment of the stylopharyngeus and stylomandibular ligaments.

Ethmoid Bone. The ethmoid bone, found immediately below and behind the frontal bone, is spongy in composition, and is very irregular in shape. Basically, it consists of a cribriform (horizontal) plate, a median perpendicular plate, like the stem of a T, and two lateral masses, or *labyrinths,* suspended from the ends of the cribriform plate. The cribriform plate helps to complete the floor of the anterior cranial fossa. The perpendicular plate forms part of the nasal septum. On either side the labyrinths form part of the orbit and the lateral wall of the nasal cavity. In addition, the inner face of the lateral mass bears a thin curving sheet forming the middle and superior nasal turbinates. Between these two lies the superior meatus; the middle meatus lies beneath the middle turbinate. The labyrinths also contain the ethmoid air cells, or ethmoid sinuses, which communicate with the nasal cavity.

Sphenoid Bone. The sphenoid bone is often likened in appearance to a bat or a butterfly with outspread wings; it also has many characteristics of a figure from an inkblot test, and requires almost as many words to interpret. It is situated at the anterior part of the base of the skull, just posterior to the ethmoid and immediately in front of the temporal, and binds the other cranial bones together. The relatively large body is shaped somewhat like a cube, with a great and a small wing projecting from each side, the small wings lying superior to the great wings. At the base of the great wings, the two pterygoid processes project downward, flanking the posterior opening of the nose; the lower end of each process is divided into two flat sheets of bone, the medial and lateral pterygoid plates or laminae. The lower end of the medial plate, the longer and narrower of the two, ends in a hook-like projection, the hamulus of the pterygoid.

The body of the sphenoid bone is hollowed into two separate spaces, the sphenoid sinuses. On the upper surface superior to the sinuses is a small fossa or pit, the sella turcica, which cradles the hypophysis (pituitary gland).

The lateral pterygoid plate is broad, thin, and flaring; its lateral surface gives attachment to the external pterygoid muscle, while its medial surfaces provides attachment for the internal pterygoid. The medial pterygoid plate, along the full length of its posterior edge, serves as attachment for the pharyngeal aponeurosis; the superior constrictor muscle takes origin from its lower third.

At the divergence of the lateral and medial plates a V-shaped space is created, the pterygoid fossa. Above this fossa is a small oval depression, the scaphoid fossa, which gives origin to the tensor veli palatine muscle; the tendon of the tensor veli palatine glides around the humulus.

The Facial Skeleton

Fourteen bones surround the nose and mouth and comprise the facial skeleton—six paired and two unpaired. Some will be discussed briefly here.

The *nasal bones* are two small oblong bones placed side by side, forming by their junction the bridge of the nose. The *zygomatic (malar) bones* arise from the midportion of the zygomatic arch and an upward projection forms part of the floor and lateral wall of the orbit. The *inferior nasal concha* (inferior turbinate) is a shell-like scroll of spongy bone that projects medially into the nasal cavity, separating the inferior from the middle meatus. It articulates principally with the maxilla. By mentioning that the *lacrimal bones* form part of the medial wall of the orbit, we have disposed of eight of the facial bones.

Vomer Bone. The vomer (plowshare) is one of the unpaired bones. It is a midline structure whose long, sloping, anterior border forms the lower and posterior part of the nasal septum. It is frequently deflected to one side at its anterior end, resulting in a deviated septum and consequent interference with nasal breathing. Its upper border thickens and then divides into two alae for articulation with the inferior surface of the sphenoid. Its posterior border is free and forms the wall between the posterior nares.

Palatine Bones. The palatine bones are located at the posterior end of the nasal cavity, behind the maxilla and in front of the medial pterygoid plate of the sphenoid.

The palatine bones are L-shaped, with one reversed so that they join at the tips of the lower (horizontal) arms, their union forming the nasal crest on the superior edge and the posterior nasal spine on the posterior border. The posterior nasal spine gives attachment to the muscle of the uvula. The superior surface of the horizontal part forms the back part of the floor of the nasal cavity, while the inferior surface forms, with the corresponding surface of its mate, the posterior fourth of the hard palate. The posterior border is free, and serves for the attachment of the soft palate.

The vertical portion forms part of the lateral walls of the nose, and on this surface bears three shallow depressions corresponding to the three nasal meatuses.

Maxillary Bones. Some problem in conceptualization arises here, since many of us were originally taught to say "maxilla" rather than "maxillae." We have referred to this bone in previous chapters as singular, and will continue to do so as soon as the exigencies of the present chapter are behind us. For the sake of accuracy we must report, however, that the maxillae are really two, twin bones so deftly joined, in most cases, as to present a deceptively unitary appearance. The exception, of course, is found in cases of cleft palate, a failure of the two bones to fuse normally during embryonic development. The combined maxillae are exceeded in size only by the mandible among the bones of the face. The mandible also represents the fusion of two halves (the symphysis menti) but remains in unwed solitude among the pages of descriptive atlases.

Each maxilla (right and left) consists of a body and four processes: the *frontal, zygomatic, alveolar,* and *palatine* processes. The body has a pyramidal shape, and contains the large maxillary sinus which follows the same form. The sinus cavity extends laterally over the maxillary molars; at times only a thin layer of bone lies between the floor of the sinus and the apices of the molar roots. In rare cases no bone at all separate the apices and the sinus, only the soft tissue of the periodontal membrane on the tooth and the mucous membrane lining the sinus.

The frontal process forms the lateral wall of the lower part of the nasal cavity. The zygomatic process is the lateral projection that articulates with the zygomatic bone and forms the anterior end of the zygomatic arch. The alveolar process, thick and spongy, is the dental arch. The palatine process is the horizontal shelf of bone within the dental arch which forms the greater portion of the floor of the nose and about three-fourths of the hard palate. As we have seen, the horizontal plate of the palatine bone supplies the other fourth.

The Mandible. The mandible consists of a horseshoe-shaped body and a strong ramus extending upward and slightly posteriorly from each free end of the horseshoe. The fusion of the two halves of the mandible is at the *symphysis menti,* which is elevated to form the mental proturberance at the point of the chin. The surface of the mandible is marked by two lines; externally, the *oblique line* is a slight ridge running downward and forward from the anterior border of the ramus, while internally the *mylohyoid line* is a marked horizontal ridge below the alveolar process.

In the upper portion of the ramus is the *mandibular foramen,* through which pass the nerves and arteries for the lower teeth. Overhanging the foramen on the inner surface is a spine of bone called the *lingula.* The *mylohyoid groove* extends downward and forward from the foramen, and is distinct from the mylohyoid line described above.

Each ramus ends in two prominent extensions, the *coronoid* and the *condyloid processes.* The coronoid is the anterior process to which attach the temporalis and masseter muscles. The condyloid process consists of a condyle, or head itself, and a neck from which the head arises. The head articulates with the mandibular fossa of the temporal bone, forming the only freely moving joint in the skull, the temporomandibular joint. Between the two processes is a broad deep depression called the *sigmoid notch,* while inferiorly the rami join the main body at a location known posteriorly as the *angle* of the mandible.

The Temporomandibular Joint. The TMJ may be palpated by placing a finger immediately forward of the ear, then opening and closing the mouth. A clearer concept may be gained by placing a fingertip inside the ear, then opening and also moving the mandible from side to side.

Shown in Figure 5-2 are three elements composing this joint: (1) the mandibular condyle; (2) the articular fossa (the glenoid fossa) of the temporal bone; and (3) the articular meniscus, often referred to simply as the *disc.*

The knuckle-like head of the condyloid process has a thicker lateral dimension than is apparent in head films or lateral drawings. It swells up in all directions from the neck, providing a relatively broad expanse of slick, cartilaginous, articulatory surface.

The articular fossa may also be somewhat deceiving in structure, since it is partially concealed behind the zygomatic process of the temporal bone in lateral view. The bone surrounding the rim of the fossa is fairly thick, but as it domes up to the roof of the fossa it becomes quite thin centrally, where it separates the disc and the condylar head from the middle cranial fossa, the lodgment of the temporal lobe of the brain. The front portion of the fossal rim is known as the *anterior articular eminence,* or *articular tubercle,* or *eminentia.* Posteriorly, the fossa is bounded by a bony ridge, the *postglenoid process.*

The disc, or articular meniscus, is a body of dense fibrous cartilage lying between the head of the condyle and the articular fossa, extending somewhat forward under the anterior eminence. This oval plate is thinner at the center than around the edges, with an undersurface concave in shape, conforming to the head of the condyle. However, anteroposteriorly

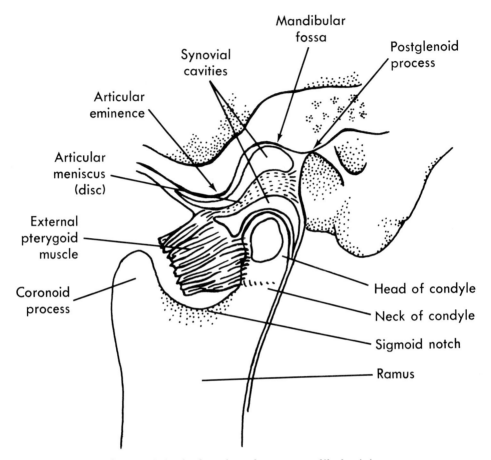

Figure 5-2. Sagittal section of tempomandibular joint.

its upper exterior has a double curve; the posterior portion is convex front-to-back, adapting to the fossa, while the anterior portion is concave, dipping under the articular eminence.

The joint is encased in a fibrous *capsule,* a sheath or sac of tissue that is attached superiorly around the circumference of the glenoid fossa; below, it attaches around the neck of the condyle. The capsule is composed of two layers, an exterior of fibrous tissue reinforced by accessory ligaments, and an inner layer, the synovial membrane, that secretes the lubricating fluid, *synovia,* into chambers above and below the disc.

The meniscus is attached medially and laterally to the neck of the condyle, so that the disc is carried with the condyle in some movements. Posteriorly it attaches to a thick layer of loose, vascular connective tissue which fuses with the capsule but which allows anterior movement of the disc without stress. Anteriorly, the disc is fused to the capsule; fibers of the external pterygoid muscle penetrate the capsule and insert into the disc at its capsular attachment, enabling the muscle to pull both meniscus and condyle forward in unison.

The surface of the thin dome forming the top of the glenoid fossa is devoid of cartilaginous armor, which suggests that the head of the condyle was not designed to press upward into the depths of the fossa during any aspect of function. To the contrary, the surface of the posterior slope of the eminence, like the articular surface of the condylar head, is covered by fibrous cartilage, indicating that these parts of the joint are able to provide resistance to force.

The entire complex of the TMJ is suspended from the skull by a system of three ligaments, guy wires which steady and brace while still allowing for freedom of movement within a prescribed range. The *temporomandibular* ligament connects above to the zygomatic arch, and below to the outer and posterior surface of the neck of the condyle. Medially, the mandible is anchored by the *sphenomandibular* ligament; this is attached above to the spine of the sphenoid bone, and below to the inner surface of the mandible at the lower border of the mandibular foramen. The stylomandibular ligament is the posterior support; it extends from the styloid process of the temporal bone to the posterior border and angle of the mandible.

Adjacent Bones

Three other bones must be considered that are not part of the skull— the hyoid bone, because of its great importance to the muscles with which we deal, and the first two vertebrae because they provide helpful landmarks.

Hyoid Bone. The hyoid bone has been called "the skeleton of the tongue" because of its many attachments of muscles supporting the tongue. It is the only bone in the human skeleton not articulating or connecting more intimately with any other bone.

The hyoid is another horseshoe-shaped bone, lying below the mandible and roughly paralleling the body of the mandible in its orientation. It is suspended in its position at the upper border of the larynx by two thin fibrous cords attached to the styloid processes of the temporal bone, the stylohyoid ligament.

The hyoid consists of a body and two pairs of processes, the greater and lesser cornua (horns). The body is quadrilateral, with a faceted anterior surface that is somewhat convex. The posterior surface is smooth and sharply concave. The epiglottis stands within the arch of this bone, and is separated from the body of the hyoid by the hyothyroid membrane and by loose connective tissue.

From each end of the body, the greater cornua curve laterally and posteriorly, tapering somewhat before swelling to a tubercle at the tip. The lesser horns are small, cone-shaped, and project upward and backward from the superior surface of the body at or near the attachment of the greater cornua. The apex of the lesser cornua provide the inferior attachment of the stylohyoid ligament, the tether of the hyoid. In addition to ligaments, a dozen or so muscles have attachment to some portion of the hyoid bone.

Cervical Vertebrae 1 and 2. The first cervical vertebra is called the *atlas,* because it supports the globe of the head; the second vertebra is named the *axis,* because it provided the pivot on which the atlas turns.

The atlas is a ring of bone with no body and no spinous process, but with a rounded prominence rising from its anterior surface, the *tubercle of the atlas,* identifiable on most lateral head films. There is also a posterior tubercle with which we are less concerned. Nodding movements occur between the atlas and the skull.

The axis is characterized by the *dens,* a tooth-like process that projects upward from the anterior region of the body. The dens articulates within the atlas; rotating movements of the head occur around the dens.

Myology

Based on the information in the preceding pages, we are prepared at last to examine the muscles from which we derive our professional title.

General Considerations

Before looking at individual muscles a number of general factors applying to all muscle tissue should be understood. A muscle consists of a bundle of elongated cells that are called *fibers,* unified by an intercellular substance and supported by a framework of connective tissues. All motion in the body is a product of the contractility of muscle fibers.

Muscles are distinct from ligaments, tendons, and aponeuroses. The latter three are composed of strong, fibrous connective tissue, are inelastic, and do not contract, but serve as the attachment of muscles to bones or other parts. Ligaments also bind together the articular ends of bones, support certain visceral organs, and are usually sheet-like in form. A tendon is a narrow ribbon-like band; an aponeurosis is a broad flat tendon.

Muscle Classification. Muscles are classified according to their structure and function into three categories; *smooth (visceral), striated (skeletal),* and *cardiac.* Smooth muscles are those of the stomach, intestines, and blood vessels, and since they are under the control of the autonomic nervous system they are not subject to conscious voluntary action. They are called smooth muscles because their cytoplasm has no transverse striations. They are composed of spindle-shaped cells held together by fibrils.

Cardiac muscle is striated, but not distinctly so. Its cells are elongated, branching, and multinucleated, and are grouped in bundles. Its contraction ejects blood from the heart and, of course, is not under voluntary control.

The striated muscles, those with which we deal, are so called because the cytoplasm of the cells has fine transverse bands that are visible microscopically. The fibers are arranged in compact bundles separated by fine connective tissue sheaths; these sheaths extend into the muscle from an outer connective tissue framework that carries blood vessels and nerves. These are the voluntary muscles.

Attachment. In describing muscle attachments two terms are used, *origin* and *insertion.* Actually, muscles do not originate at one point and insert into another, but convention decrees that we identify the attachment nearer the center of the body, or the one more fixed, as the origin, and the more peripheral attachment, or the one more movable, as the insertion. The supposition would be that contraction of the muscle would create movement from the insertion toward the point of origin; in function, this is not always the case.

In general, muscles are attached to bones; however, some muscles are affixed to cartilage while others encircle blood vessels (vasoconstrictors) or attach to visceral organs, etc. Occasionally a muscle is anchored directly to the periosteum of bone, but more commonly the attachment is made by means of a tendon or aponeurosis.

Nerve Supply. Each muscle is supplied with at least one nerve that transmits impulses from the central nervous system (CNS), causing the muscle to contract. While each motor nerve fiber is capable of innervating up to fifty muscle fibers, those muscles that perform delicate tasks have a larger proportion of nerve fibers, ranging down to only six muscle fibers for each nerve fiber. A motor nerve fiber with the muscle fibers it serves is called a *motor unit.* Sensory nerves feed back information regarding the degree and strength of contraction to the central nervous system.

The many interconnections and interrelationships of the nervous system find some reflection in muscular tissue. No movement is the result of action by a single muscle. Many other muscles function simultaneously to a greater or lesser degree, supplementing, stabilizing, smoothing, or simply relaxing antagonistic tension, in order to execute the desired action.

Types of Contraction. Striated muscle is capable of two types of contraction, both of which should be understood by the clinician. When a muscle bundle contracts, lifting a weight, the muscle becomes shorter and thicker, but its tonus remains the same. Since the tone of the fibers is not changed, this contraction is called *isotonic.* On the other hand, if the muscle is forced to contract against a weight that it cannot lift, the tension in the fibers increases, but the muscle length is not altered. Since the length of the fibers is unchanged, this contraction is called *isometric.* Most muscle contraction is of the isotonic type, but complex muscular activity requires the coordinated development of both types.

Musculature

Of more than 600 muscles in the body, we will study approximately 40. While it will be necessary to discuss some of these individually, we will again present much of the information in chart form.

Muscles of Expression. The array of muscles customarily grouped under the heading of "muscles of facial expression" includes a number situated about the eyes, nose, and scalp. We will eliminate those as being nonessential to our specific purposes. Nevertheless, we are left with ten muscles in which we have a rather intense interest, those listed in Table 5-1. The location of each can be seen in Figure 5-3. All are innervated by cranial VII, the facial nerve.

Orbicular Oris. We may logically consider first the orbicularis oris, the most complex of the facial muscles, the only one not demonstrably paired in the two sides of the face, and the one creating most anguish for the clinician in many cases. The orbicularis oris encircles the lips and is the sphincter of the mouth, its tonicity alone usually adequate to achieve bilabial contact in the resting face. A slight increase in tension suffices to

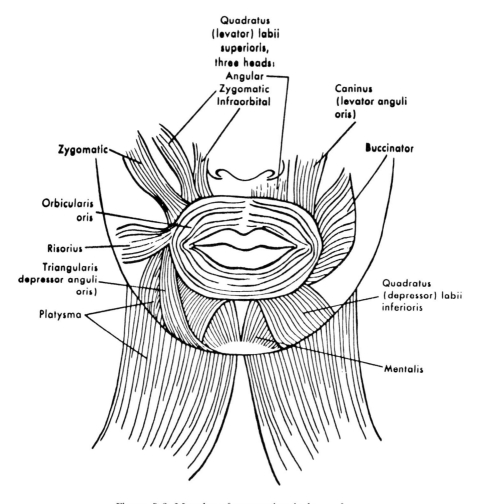

Figure 5-3. Muscles of expression in lower face.

maintain closed lips during mastication, while extreme contraction puckers the lips.

The complexity of the orbicularis oris arises partly from the fact that its fibers appear to be borrowed from other perioral muscles; technically it has no origin or insertion, and might be considered in a sense to be a composite of many muscles, a hub where representatives from surrounding muscles meet to form a unified network. As a result, this muscle is not simply a circular band. As the fibers pass along the border of the lips, there are intimate interconnections between the decussating fibers of the upper and lower lip, but many fibers then pass on at various oblique angles into the neighboring muscles. It has now been shown that contraction of orbicularis oris does not necessarily result in equal force of the upper and lower lip; one or the other lip may be weaker, a condition that would not exist if the muscle were a continuous band.

Table 5-1. Muscles of Facial Expression

Muscle	Origin	Insertion	Function
Orbicularis oris	None. Derives fibers from other muscles in the area. Encircles mouth.		Closes mouth; presses or puckers lip
Quadratus labii superior	Frontal process of maxilla; lower border of orbit; zygomatic bone	Upper lip, laterally from midline	Elevates upper lip
Caninus	Canine fossa of maxilla	Upper lip, at angle of mouth	Pulls corner of mouth up and back
Zygomatic	Zygomatic bone	Upper lip, at angle of mouth	Pulls corner of mouth up and back
Buccinator	Alveolar processes of maxilla and mandible; pterygo-mandibular raphe	Fibers of other labial muscles	Compresses cheek: unifies action of interconnecting muscles
Risorius	Fascia over masseter	Skin at angle of mouth	Retracts angle of mouth
Triangularis	Oblique line of mandible	Lower lip at angle of mouth	Depresses angle of mouth
Quadratus labii inferior	Oblique line of mandible	Integument of lower lip	Draws lower lip down and back
Mentalis	Incisive fossa of mandible	Integument of chin	Raises and protrudes lower lip
Platysma	Fascia covering thoracic muscles	Skin and tissue of lower face; mandible below oblique line	Depresses mandible; wrinkles skin of neck and chin; depresses outer end of lower lip

Two minor muscles that we will not pursue hereafter serve to stabilize the orbicularis oris by providing bony attachment. The upper one consists of two bands, lateral and medial, on either side of the midline; the lateral band connects the orbicularis with the alveolar border of the maxilla near the lateral incisor tooth, while the medial band connects to the septum of the nose. The interval between the two medial bans is recognized on the surface of the lip as the *philtrum,* the vertical depression beneath the nasal septum terminating at the vermillion line. The lower lip is attached to the mandible, lateral to the mentalis, by the second of these two muscle slips.

We will now move clockwise around the orbicular oris, although the overlapping of muscles will preclude their precise delineations.

Quadratus Labii Superior. The quadratus labii superior (levator labii superioris) lies nearest the midline of the muscles above the mouth, some of its fibers even inserting into the alar cartilage of the nose. It has three heads: the *angular,* whose origin is the frontal process of the maxilla near the nose, the *infraorbital,* with origin at the lower border of the orbit, and the *zygomatic,* arising from the zygomatic bone. All three run inferiorly and insert into the upper lip from the midline laterally.

Caninus. The caninus has origin in the canine fossa below the infraorbital foramen of the maxilla, and thus lies quite near the superior quadratus. Its fibers descend vertically to the angle of the mouth, its insertion in the upper lip mingling with fibers from the orbicularis, triangularis, and zygomatic muscles.

Zygomatic. The origin of the zygomatic muscle is from the zygomatic bone, placing it lateral to the caninus. Its fibers descend obliquely with a medial inclination and insert into the corner of the mouth, blending with fibers of the nearby muscles.

Buccinator. The buccinator covers the entire lateral wall of the mouth and constitutes the essential muscular coat of the cheek; it will be discussed in more detail in the following chapter, when we explore its function. It arises in the molar region from the outer surfaces of the alveolar processes of the maxilla and mandible, and posteriorly from the anterior border of the pterygomandibular raphe, a dense band of deep fascia in the pharynx.

Buccinator fibers converge as they approach the angle of the mouth. The central fibers intersect and pass on, those from the upper portion dipping down to become continuous with the fibers of the orbicularis oris in the lower lip, lower central fibers rising to continue around the upper lip. Fibers from the upper and lower regions of the buccinator continue forward into the corresponding lip without decussation.

In basic function, contraction of the buccinator compresses the cheek, keeping it firmly in contact with the teeth, and thus aids in chewing by preventing food from escaping into the buccal vestibule.

Risorius. The risorius is a narrow bundle of fibers originating in the fascia overlying the masseter. It runs horizontally forward, tapering somewhat before it inserts into the skin at the angle of the mouth.

Triangularis. We now dip below the horizontal midline to the triangularis muscle. This muscle has origin in the oblique line of the mandible, where it is continuous with the fibers of the platysma muscle. From this broad origin the fibers converge into a narrow bundle and insert into the angle of the mouth.

Quadratus Labii Inferior. The quadratus labii inferior (depressor labii inferioris) is a small square muscle also arising from the oblique line of the mandible, but anterior to the traingularis. Its fibers run upward and medialward to insert into the integument (skin) of the lower lip where it also blends with the orbicularis oris.

Mentalis. In our clockwise journey we have now arrived at 6 o'clock, the lower vertical midline, where we find the mentalis muscle, the only facial muscle with fibers running *away* from the lips. Its origin is the incisive fossa of the mandible, from where it descends alongside the frenum of the lower lip to insert into the skin covering the chin. It is often described in anatomy texts without conscious irony as a "small" muscle; the oral myologist battles to reduce the seemingly mountainous bundle that results from excessive use. Even when not overdeveloped, it varies in size more than any of the other facial muscles.

Muscles of Mastication. The four muscles itemized in Table 5-2 are termed the "muscles of mastication," although a number of other muscles have accessory functions in varying degrees during actual mastication. A detailed description of the masticatory process was given above in Chapter 1.

Of the four muscles under discussion here, three (the masseter, temporal, and internal pterygoid) serve to elevate the jaw, and are often grouped as the "antigravity muscles." The fourth, the external pterygoid, lowers the mandible. All are supplied from the motor part of the mandibular branch of cranial V, the trigeminal nerve. This branch is sometimes referred to as the "masticator nerve."

Masseter. The masseter is a thick, powerful muscle consisting of two parts, a superficial and a deep portion, as shown in Figure 5-4, A. The superficial portion arises as a thick aponeurosis from the zygomatic process of the maxilla and the lower border of the zygomatic arch. These fibers run downward and backward, and inset into the angle of the mandible and into the lower half of the outer surface of the ramus. The deep portion, smaller but more muscular, arises from the medial surface of the zygomatic arch; its fibers pass downward and forward, where they

Table 5-2. Muscles of Mastication

Muscle	Origin	Insertion	Function
Masseter	Superficial: lower border of zygomatic arch Deep: medial surface of zygomatic arch	Superficial: outer surface of lower ramus and angle of mandible Deep: outer surface of upper ramus and coronoid process	Lifts mandible vertically
Temporalis	Temporal fossa of skull	Coronoid process and anterior margin of upper ramus	Elevates and retracts mandible
Internal pterygoid	Lateral pterygoid plate; palatine and maxillary bones	Lower margin of inner surface of ramus	Elevates and protrudes mandible
External pterygoid	Upper head: greater wing of sphenoid Lower head: outer surface of lateral pterygoid plate	Neck of condyle and articular disc of TMJ	Draws mandible forward; depresses mandible; moves mandible to side

insert into the upper half of the ramus and the outer surface of the coronoid process. Contraction of the masseter lifts the mandible vertically against the maxilla.

Temporalis. The temporalis, also seen in Figure 5-4, A, is a fan-shaped muscle whose origin covers the entire temporal fossa. As the fibers descend, they converge to end in a tendon which passes beneath the zygomatic arch and inserts into the coronoid process and the anterior border of the ramus of the mandible. From this attachment the temporalis is enabled to retract the mandible as it elevates.

Internal (medial) Pterygoid. The internal pterygoid (Fig. 5-4, B) is another thick, quadrangular muscle. It arises from the lateral pterygoid plate and from the nearby surfaces of the palatine and maxillary bones. Its fibers angle downward, backward, and outward to find insertion by a strong sheet of tendon into the inner surface of the ramus along its lower margin. In this posture it acts to pull the mandible forward.

External (lateral) Pterygoid. The external pterygoid is a short, thick muscle that tapers backward from twin origins almost horizontally to the condyle of the mandible. It, too, is shown in Figure 5-4, B. One of its heads

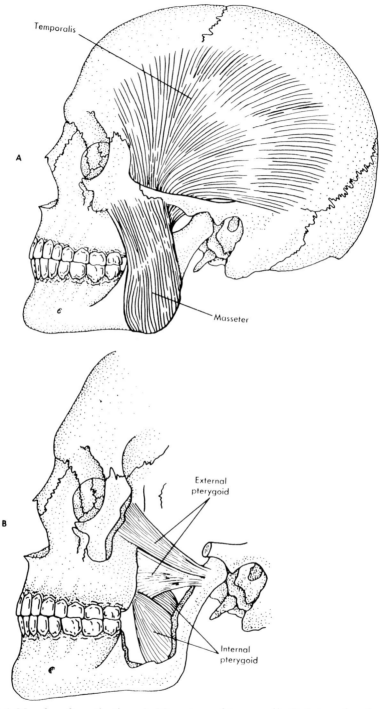

Figure 5-4. Muscles of mastication. A, Masseter and temporalis. B, Internal and external pterygoids.

arises from the outer surface of the greater wing of the sphenoid bone, the other from the outer surface of the lateral pterygoid plate of the sphenoid bone. Its fibers incline slightly outward as they pass back to their insertion on the front of the neck of the condyle. Some fibers also insert into the front margin of the articular disc of the temporomandibular joint. The action of the external pterygoid is in part to tip the mandible down, opening the jaw, but its primary function is to draw the mandible forward. This protrusion of the mandible results when the external pterygoid of both sides contract in unison. When both the external and internal pterygoids of one side work alternately with those of the other side, they produce lateral movement of the mandible.

Muscles of the Tongue. The tongue, the target of much of the oral myologist's efforts, is composed almost entirely of muscles. It must then be obviously essential that the clinician know and understand the eight constituent muscles of the tongue. Four of these are contained entirely within the tongue, the *intrinsic* muscles, and are responsible for changes in the shape of the tongue; four others originate from nearby skeletal structures, thus are *extrinsic* muscles, and are accountable for most movements of the tongue.

The tongue is divided into lateral halves by a median fibrous septum that extends the entire length of the tongue. This division is outlined on the surface, beneath by the lingual frenum, the vertical fold of mucous membrane along the midline that attaches the tongue to the soft tissue below, and above by the midline groove running front to back on the dorsum of most tongues. All lingual muscles are thus paired, each half of the tongue boasting a full set of eight.

Intrinsic Muscles. The intrinsic muscles of the tongue are displayed in Table 5-3. They are named for their planes of direction within the tongue and, although there is considerable interweaving of fibers, they remain to a great degree separate bundles or layers. Fibers from the extrinsic muscles also blend in, providing the tongue with the potential for an infinite versatility while maintaining firm coordination between the position and the shape of the tongue.

We need add little here to the information in Table 5-3. The *superior longitudinal* is a thin layer laying immediately beneath the mucous membrane on the dorsum of the tongue. The *inferior longitudinal* is a narrow band on the under surface of the tongue, between the genioglossus and hyoglossus. The fibers of the *transverse* pass laterally from the midline to the lingual edges. The *vertical* is found only at the borders of the forepart

Table 5-3. Intrinsic Muscles of the Tongue

Muscle	Origin	Insertion	Function
Superior longitudinal	Submucosa near epiglottis; median septum of tongue	Edges of tongue	Shorten, widens tongue; turns tip and sides up, forming concave dorsum
Inferior longitudinal	Lower portion of root of tongue	Apex (tip) of tongue	Shortens, widens tongue; depresses tip, forming convex dorsum
Transverse	Median septum of tongue	Mucosa at sides of tongue	Narrows, elongates tongue
Vertical	Upper surface of tongue	Lower surface of tongue	Flatten, widens tongue tip

of the tongue. All are innervated by cranial X11, the hypoglossal nerve.

Extrinsic Muscles. The four extrinsic muscles of the tongue are shown in Table 5-4 and pictured in Figure 5-5. The palatoglossus muscle is included, since it is a lingual muscle; however, it is also associated with the soft palate and its function may be best understood in that context. Accordingly, it will be described in the following section.

Genioglossus. The genioglossus muscle derives its name from the Latin *genion* (chin) and the Greek word for tongue (*glossa*); the Latin term for tongue is *lingua.* This muscle not only constitutes a large physical proportion of the tongue, it does most of the work. It also poses one of the major challenges for the clinician, since lack of a normal balance of tonus among the three segments of the genioglossal fibers allows the tongue to drift forward and downward to a familiar but undesirable interdental resting posture.

The genioglossus is a flat, triangular muscle, somewhat resembling a fan held perpendicularly, the handle representing its point of origin. This origin is a short tendon from the lingual surface of the mandible at the symphysis, from where the fibers fan out in three directions. The lower fibers extend back and down where they attach to the upper part of the body of the hyoid bone. The middle fibers pass back and upward, while the superior ones curve up and forward; combined, they enter the entire length of the tongue from root to tip to insert along the median fibrous septum.

The anterior fibers withdraw and depress the tip of the tongue. They also work in concert with the middle fibers to draw the entire dorsum of

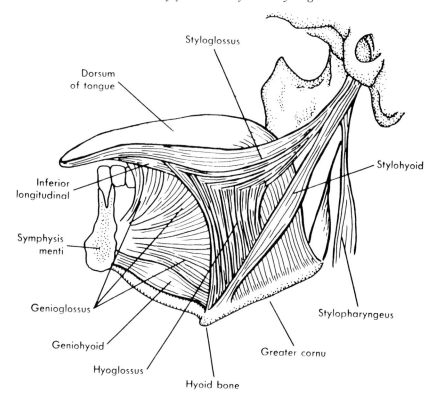

Figure 5-5. Extrinsic muscles of the tongue.

the tongue downward into a concave channel, as in sucking action. The middle portion alone draws the base of the tongue forward and thereby protrudes it. The inferior fibers pull the hyoid bone upward and forward, a part of the synergy of actions involved as the entire larynx moves up and forward during deglutition.

Hyoglossus. The hyoglossus is a thin, rectangular sheet of muscle that arises from the side of the body of the hyoid bone and from the entire length of its greater cornu. It runs vertically upward to enter the side of the tongue, then passes medially to insert into the medial septum of the tongue. Its fibers interlace with intrinsic muscle fibers. Its contraction pulls the tongue down and back; when the hyoid bone is fixed, it depresses the sides of the tongue.

Styloglossus. The styloglossus originates from the styloid process of the temporal bone. As it passes downward and forward to the side of the tongue, it divides into two segments, the longitudinal and oblique portions. The longitudinal portion enters the side of the tongue near its dorsal

Table 5-4. Extrinsic Muscles of the Tongue

Muscle	Origin	Insertion	Function
Genioglossus	Inner surface of mandible near symphysis	Median septum of tongue; body of hyoid bone	Various fibers protrude and retract tongue; depress midline of tongue; elevate hyoid bone
Hyoglossus	Body and greater cornu of hyoid bone	Side of the tongue, posterior half	Depresses and retracts tongue; depresses side of tongue
Styloglossus	Styloid process of temporal bone	Lateral margin, full length of tongue	Draws tongue upward and backward; elevates side of tongue
Palatoglossus	Anterior surface of velum	Side of posterior tongue	Constricts faucial isthmus; elevates posterior tongue

surface, and blending with the fibers of the inferior longitudinal muscle, runs to the tip of the tongue. The oblique portion overlaps and penetrates the hyoglossus muscle as it runs transversely to the midline of the tongue. The styloglossus retracts the tongue and draws its sides upward.

Muscles of the Soft Palate. Looking over the tongue toward the back of the mouth, as in Figure 5-6, several features are to be noted. The soft palate (velum) extends from its junction with the hard palate without perceptible demarcation, the continuous mucous membrane concealing the line of the union. The attachment of the velum to the hard palate is achieved by the *palatine aponeurosis,* a thin, firm, fibrous plate emerging from the posterior border of the hard palate. When lifted, the velum reveals the uvula suspended from its lower midpoint, as well as the posterior pharyngeal wall behind. Laterally the soft palate blends into the two *pillars of the fauces* (throat), one behind the other, which connect the velum with the lateral pharyngeal wall and the base of the tongue, respectively. The posterior pillar is called the *palatopharyngeal arch* and the anterior is termed the *palatoglossal arch;* each is formed by a muscle described below. Between the two arches on either side lie the tonsils, technically the *palatine tonsils.* The lateral space between the anterior faucial pillars, the opening into the pharynx, is called the *faucial isthmus.*

We may then examine the velum more closely. This mobile structure,

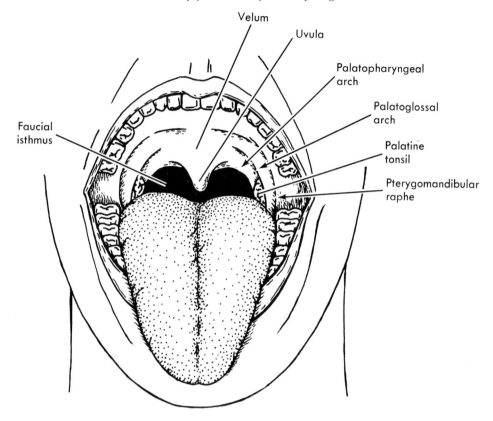

Figure 5-6. Posterior area of mouth.

like the tongue, is largely composed of muscles. Because these muscles enter the palate from various outside locations, some above, some below, and some behind the velum, it is given the ability to move in many directions as it combines or separates the oral and nasal cavities, the nasal cavity and the pharynx, etc. The five muscles thus employed are itemized in Table 5-5, and some are depicted more graphically in Figure 5-7.

Note that we enter here into a gray area where voluntary contraction of muscles begins to give way to reflex contraction; posterior to the velum, direct voluntary control over muscle is lost, contraction occurring only as a secondary reaction to some volitional action in the mouth. That is, we cannot deliberately set in motion the chain of constrictor muscles lining the pharyngeal wall; we can only perform a voluntary act that will trigger their reflexive contraction.

The importance of the foregoing lies in the fact that, among patients we may see, many display inactivity, or inappropriate activity,

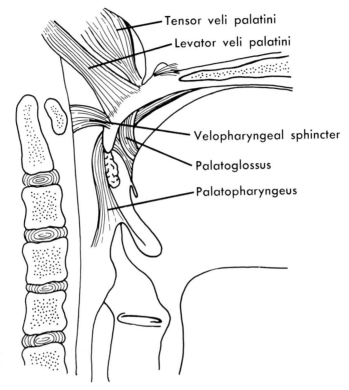

Figure 5-7. Muscles of soft palate.

of the palatal muscles. While voluntary control of these muscles may not be exercised, it is accessible, and needs to be established in most cases.

The muscle of the *uvula* is not one to cause concern. It provides some help to the levator veli palatini in achieving velopharyngeal closures.

The *tensor veli palatini* and the *levator veli palatini* need little discussion beyond that shown in Table 5-5. Both are named for their function.

The *palatopharyngeus,* as indicated above, forms the posterior faucial pillar. Its upper and lower fasciculi unite into a compact bundle to curve out and down behind the tonsil.

The *palatoglossus* forms the other (anterior) pillar of the fauces, arching down in front of the tonsil. Since both ends attach to mobile structures, its contraction may either draw the sides of the tongue up and back, or draw down the sides of the velum, depending upon which end of the muscle is more firmly fixed at the moment.

Muscles of the Neck. Moving dorsad, we have now reached the outer boundary of our province; as previously noted, the sequential layer of

Table 5-5. Muscles of the Soft Palate

Muscle	Origin	Insertion	Function
Uvula	Posterior nasal spine; palatal aponeurosis	Body of uvula	Elevates uvula
Tensor veli palatini	Scaphoid fossa; spine of medial pterygoid plate; posterior border of hard palate	Palatal aponeurosis; cartilage of the eustachian tube	Tenses velum; opens eustachian tube during deglutition
Levator veli palatini	Lower surface, petrous portion of temporal bone; side of eustachian tube	Throughout velum to midline	Elevates velum toward posterior pharyngeal wall; dilates oriface of eustachian tube
Palato-pharyngius	Lower: mucous membrane along posterior border velum Upper: midline of velum	Posterior thyroid cartilage; aponeurosis of pharynx	Depresses velum; constricts faucial isthmus; elevates pharynx
Palatoglossus (see also: "Extrinsic muscles of tongue")	Anterior surface of velum	Side of posterior tongue	Constricts faucial isthmus; elevates tongue

muscle lining the wall of the pharynx is beyond our supervision. We have described two muscles that enter this region, the palatoglossal and palatopharyngeal. Five other muscles reside in the area, the *superior, middle,* and *inferior constrictor,* aided from below by the *stylopharyngeal* and *salpingopharyngeal* muscles. These function in series to change the shape of the tube in deglutition, constricting the diameter of the pharynx, as the bolus is progressively squeezed into the esophagus.

Inferior to the tongue there are four additional muscles in which the clinician should have an interest, those treated in Table 5-6. These all lie above the hyoid bone, to which each one is attached, and accordingly they are known as the suprahyoids. Four other muscles, the infrahyoids, attach to the hyoid bone from below; as with the constrictors, the infrahyoids are beyond our grasp on a direct basis. The infrahyoids consist of the *sternohyoid, sternothyroid, thyrohyoid* and *omohyoid* muscles, and serve at the conclusion of the swallowing act to return the hyoid bone and the larynx to a neutral position.

The *digastric* muscle is a fleshy, sling-like muscle consisting of two bellies arising from opposite points but joining near the middle. The anterior belly originates on the inner and lower border of the mandible close to the symphysis and loops downward and backward. The posterior belly arises from the mastoid notch of the temporal bone and swings downward and forward. The two bellies meet at a round intermediate tendon which in turn gives rise to the *suprahyoid aponeurosis,* a broad fibrous layer that attaches to the body and greater cornu of the hyoid bone. It is somewhat unusual in that the two bellies are not served by the same nerve, the anterior being innervated by the trigeminal, cranial V, while the posterior belly receives fibers of the facial nerve, cranial VII.

Table 5-6. Suprahyoid Muscles

Muscle	Origin	Insertion	Function
Digastric Anterior belly	Interior mandible near midline	Body and greater cornu of hyoid bone, via intermediate tendon	Depresses mandible or elevates hyoid bone
Posterior belly	Inner surfaces of mastoid process		
Mylohyoid	Mylohyoid ridge of mandible	Body of hyoid bone and midline raphe	Lifts hyoid bone and base of tongue up and forward
Geniohyoid	Inferior mental spine at symphysis of mandible	Body of hyoid bone, anterior surface	Draws hyoid bone and base of tongue forward
Stylohyoid	Styloid process of temporal bone	Body of hyoid near greater cornu	Lifts hyoid bone up and backward

The *mylohyoid* has added interest because it forms the muscular sheet that is the floor of the mouth. It arises from the entire length of the mylohyoid line of the mandible, from the symphysis menti to the third molar region. The posterior fibers run medially and slightly down to insert into the body of the hyoid bone. The middle and anterior fibers meet their fellows from the opposite side at a midline raphe which runs from the symphysis to the hyoid bone.

The *geniohyoid* and the *stylohyoid* muscles work from opposite antero-posterior origins to move the hyoid bone forward and backward.

A GLOSSARY OF ANATOMIC TERMS

abduction. To draw away from the midline.

adduction. To draw toward the midline.

adipose. Pertaining to various tissues which store fat cells.

afferent. That which conducts toward the center, usually related to sensory nerves.

ala;;pl. alae. A structure or appendage which has wing-like characteristics.

amorphous. Shapeless; without distinct form or structure.

ansa. Any loop-like structure of bone or nerve.

antrum. Any nearly closed hollow or cavity, usually applied to those cavities or sinuses located in the various bones.

apex. The topmost part of a structure which is conical or pyramidal in shape; tip of the tongue.

aponeurosis. A broad, flat sheet of connective tissue serving for the attachment of muscular fibers.

areolar. Related to the fleecy, mesh-like organization of connective tissue which occupies the interspaces of the body.

articular. Pertaining to surfaces or structures which meet to form a joint.

astrocytes. Nerve cells or bone corpuscles which are star-shaped.

atlas. The first cervical vertebra.

atrium. Chamber, sinus, or cavity in the heart, lung, ear, and larynx.

atrophy. A wasting of any part due to lack of nutrition; caused by disuse, injury, disease, or interference with blood or nerve supply.

auricular. Related to the auricles of the heart, or of the ear, or to the external ear in general.

autonomic. A self-controlling structure or system; usually related to a portion of the nervous system uncontrolled by the brain or spinal cord.

axis. The second cervical vertebra; also the pivot-point, the center, or line running through the center.

bolus. A rounded or pill-shaped mass; a mass of food ready to be swallowed.

bronchiole. Name applied to each of the divisions resulting from the forking of the trachea.

buccinator. The trumpeter; a flat muscle located in the cheek, which helps to compress the cheeks.

cancellous. Having a latticework structure, as the tissue of spongy bone.

capitulum. A small, rounded protuberance on a bone surface.

capsule. An enveloping membrane acting as a container.

cardiac. Pertaining to the heart, or esophageal orifice of the stomach.

carotid. The artery which runs to the brain, or to the head in general; the principal artery of the neck.

cartilage. A dense, nonvascular body tissue capable of withstanding considerable pressure or tension; it furnishes strength, shape, and flexibility.

caudad. Toward the tail or lower end.

caudate. Possessing a tail.

cephalad. Toward the head or upper end.

cervical. Related to the region of the cervix or neck.

choana. Posterior opening into the nasopharynx of the nasal cavity.

cilia. Hair-like processes projecting from epithelial cells. They line the respiratory tract, waft only in one direction, and wave mucus, pus, and dust particles upward from the bronchi and backward from the nasal passage.

coccygeal. Pertaining to the last four bones of the spine.

collagen. An insoluble fibrous protein, the chief constituent of the fibrils of connective tissues. It is characterized by swelling in water solutions, by conversion to gelatin and glue on prolonged heating in water, and by conversion to leather on tanning.

collagenous. Containing or composed of collagen.

concha. A rounded protuberance at the end of a bone forming an articulation, as on the mandible.

contralateral. Situated on, pertaining to, or affecting the opposite side, as opposed to ipsilateral.

cornu (plural, **cornua**). A structure shaped like a horn.

coronal. Any structure resembling a crown; the anatomical plane which divides the body into front and back sections.

corpus. Any mass or body.

cortex. Any outer layer of substance; usually applied to the outer layer of the brain (cerebral cortex) or of bone.

cranial. Related to the head or upper end.

cribriform. Perforated like a sieve.

cricoid. Signet-shaped; the cricoid cartilage in the larynx which acts to support the thyroid and arytenoid cartilage.

cuneiform. A wedge-shaped structure; cartilages located near the arytenoid cartilages of the larynx.

cutaneous. Pertaining to the skin.

cytology. The study of cellular structures.

cytoplasm. Protoplasm; cell plasm lying external to the nucleus.

decussate. A crossing action which results in an X formation; usually applied to nerve groups of the central nervous system.

deferens. That which carries away from the center; usually applied to nerves or ducts.

deglutition. The act of swallowing.

diaphysis. The portion of a growing bone called the shaft.

diarthrosis. A freely moving joint such as the elbow, wrist, temporo-mandibular, etc. (also, **diarthrodial**).

diagastric. Double-bellied muscle of mastication located near the floor of the mouth.

distal. Away from the point of attachment, away from the midline.

dorsal. Pertaining to the back or rear side.

ectoplasm. The outer layer of the protoplasm which forms a cell membrane.

edema. An abnormal swelling owing to an accumulation of serum in a tissue or part.

efferent. That which conveys away from the center; usually applied to motor nerves.

embryology. The study of cell growth and the development of an organism.

endo-. Prefix meaning within, or inner.

enervate. To make weak; to lessen the nerve, vitality or strength of.

epi-. Prefix meaning upon, above, or upper.

epiphysis. The ends of a growing bone which form the boundaries for the diaphysis or shaft.

epithelium. The cellular, outer substance of skin and mucous membrane which is without blood supply.

esophagus. Part of the digestive system; the tube connecting the pharynx and the stomach.

eustachian tube. The tube which leads from the middle ear to the pharynx; also known as **auditory tube**.

extrinsic. Derived from or situated without; external.

falciform. Sickle-shaped.

fascia. A tough band or sheet of fibrous connective tissue that encloses, supports, and separates muscles and some organs.

fasciculus. A bundle or cluster of nerve or muscle fibers which gather to make up whole nerves or muscles.

fauces. The space leading from the mouth into the pharynx.

fenestra. Window or opening in the middle ear that leads to the inner ear.

fibril. A small fiber; a very small filament, often the component of a cell or fiber.

fissure. A groove or sulcus; usually applied to those clefts found in tissues.

fixate. The act of making static, fixed, or relatively immovable.

flaccid. Soft, flabby, usually applied to muscles which have lost their quality or tonus.

follicle. A small sac or gland capable of excretion.

foramen. An opening or passage that is a regular part of a structure.

fossa. A furrow, cavity, or depression that is a regular part of a structure.

fusiform. Having a spindle shape.

glossal. Pertaining to the tongue.

glossodynia. Neuralgic pain in the tongue.

glottis. The opening between the vocal folds which disappears when the folds approximate.

hamulus. Any hook-shaped structure; usually applied to the pterygoid process of the sphenoid bone.

hiatus. A space, gap, foramen, or opening in any structure.

histology. The study of tissues.

homeostasis. The state of equilibrium of the internal environment in the living body with respect to various functions.

hormone. A chemical secretion of the ductless glands which is carried in the bloodstream and which acts to stimulate the activity of organs.

hyaline. Type of white, pearly cartilage commonly found throughout the body.

hyoid. The U-shaped bone located below the tongue and above the thyroid cartilage.

hyper-. Prefix meaning in excess of some normal state.

hypo-. Prefix meaning less than some normal state.

inferior. Below; toward the caudal end of the body.

infundibulum. A funnel, canal, or extended cavity; usually applied to a passage connecting the nasal cavity with ethmoid bone or with the area at the upper end of the cochlear canal.

innervate. To stimulate a part, as the nerve supply of an organ.

insertion. Point of attachment for muscles; usually applied to the most movable point of attachment.

integument. A covering; the outermost surface of the body, or skin.

interstitial. Located in the spaces between cells, or between essential parts of an organ.

intrinsic. Anything wholly contained within another structure; usually applies to muscles which are exclusively attached to one organ or structure, as intrinsic muscles of the tongue.

ipsilateral. Situated on the same side.

jugular. Pertaining to the neck; usually applied to the large vein of the neck.

lacrimal. Pertaining to the tears and the ducts from which they arise.

lacuna. A small, hollow space; usually applied to bone cavities.

lamina. A plate or flat layer of bone.

larynx. The structure responsible for voice; the enlarged upper end of the trachea; the voicebox.

lateral. Away from the midline, toward the periphery.

ligament. Tough, fibrous connective bands which support or bind bones and various organs.

lobe. A globular part of an organ usually delineated by fissures or cleavages.

lumbar. Pertaining to the portion of the back located between the thorax and the pelvis, commonly referred to as the loins.

lumen. The cross-sectional space within a tubular structure.

lymph. A clear, watery fluid secreted by the lymph glands in order to expedite the removal of waste from tissue cells.

malar. Pertaining to the cheekbones.

mandible. The bone of the lower jaw; the inferior maxilla.

mastoid. Nipple-shaped; a process of the temporal bone.

maxilla. A jawbone, especially the upper one; the superior maxilla.

meatus. Any passage or opening, usually canal-like.

median. Situated in the midline.

medial. Toward the midline or median plane.

medulla. The marrow; the inner portion of an organ, in contrast to the outer portion or cortex.

membrane. A thin sheet of tissue which sheathes or divides organs and surfaces.

meninges. Specialized membranes which encase the brain and spinal cord.

meniscus. Concavoconvex lens; an interarticular cartilage of crescent shape found in certain joints, as the temporomandibular.

metabolism. The regular chemical modifications of substances which occur in the growth and development of the body.

morphology. The study of forms and structures.

mucus. The sticky secretion which covers the membranes of many cavities and passages that are exposed to the external environment.

muscle. Specialized tissues composed of contractile cells or fibers which furnish the body with motive power.

mylohyoid. A muscle connected to the mandible, the hyoid bone, and median raphe.

myo-. A Greek combining form signifying relation to muscle; meaningful only when used as an attached prefix.

nares. The anterior openings of the nasal cavities which communicate with the external environment, i.e., the nostrils.

nasopharynx. The part of the pharynx located above the velum or soft palate.

nucleus. A small round body within every cell which acts as the functional control center; refers also to a mass of cell bodies in the brain or spinal cord.

occiput. The back part of the skull.

orbicular. Circular; muscle about an opening, as the orbicularis oris muscle encircling the mouth.

orifice. An entrance or opening into a body cavity.

origin. The relatively fixed muscular attachment.

oropharynx. The part of the pharynx located between the velum, or soft palate, and the hyoid bone.

ossicle. A small bone; often applied to the bones of the ear.

ossify. The act of becoming bone.

osteoblast. A cell concerned with the formation of bone.

osteology. The study of structure and function of bone.

ostium. A small entrance or opening.

pedicle. Stalk-like process or stem.

peduncle. A supporting part of another structure; a band running between sections of the brain.

perioral. About or surrounding the mouth.

periosteum. The fibrous sheath which covers all bones except at their articular surfaces.

peristalsis. A wave of contraction passing along a tube.

petrous. Resembling stone; relating to the petrous portion of the temporal bone.

pharyngeal. Pertaining to the pharynx or throat.

plasma. The fluid portion of the blood during circulation.

platysma. A plate; usually applied to the neck muscle connected to the mandible and the clavicle.

plexus. A collection, concentration, or network of parts of the nervous or vascular systems.

posterior. Toward the back or rear side.

protoplasm. The basic material of every living cell.

proximal. Toward the point of attachment; toward the midline.

pterygoid. Wing-shaped, alate; usually applied to two large processes of the sphenoid bone or to the muscles that arise from these.

pulmonary. Pertaining to the lungs.

ramus. A branch; usually applied to parts of nerves, vessels, or bone.

raphe. A line or seam formed by the union of two parts.

risorius. A muscle connected to facial fascia and the angle of the mouth.

rostral. Toward the head end.

sacrum. The triangular bone composed of five united vertebrae and forming the base of the vertebral column.

sagittal. Arrow-like, usually applied to the plane which divides the body into right and left portions.

sarcolemma. A membranous sheath encasing each striated muscle fiber.

scaphoid. A bone or process shaped like a small boat.

segmentation. Division into similar parts or segments.

sensory. Nerves, organs, or structures related to the process of sensation and carrying stimuli from the exterior toward the cerebrospinal system.

septum. A partition or dividing wall, such as the nasal septum.

sinus. Any cavity having a relatively narrow opening, as those located in the facial bone.

somatic. Pertaining to the body, or to structures of the body wall.

sphenoid. Wedge-shaped; a complex bone of the interior skull.

sphincter. Any muscle or combination of muscles which provides a closure for a natural body opening.

squamous. Scale-like; the upper anterior portion of the temporal bone.

sternum. The breastbone.

striated. Streaked; usually applied to a special type of muscular fiber which effects voluntary movement.

styloid. Resembling a stylus; a process of the temporal bone which furnishes attachment for muscles and ligaments.

sulcus. A fissure or groove in bone or tissues, especially of the brain.

superficial. Confined to the surface; lying on or near, or affecting only the surface or surface layers.

superior. Above; toward the head or cephalic end.

suture. A stitch or seam; the line of union in an immovable articulation, as those between skull bones.

symphysis. A line of fusion between two bones which are separate in early development, as the symphysis of the mandible.

synarthrosis. A type of joint in which the skeletal elements are united, resulting in restricted or complete lack of movement in the joint.

syndesmosis. Restricted movement of a joint because of connective tissue attachments.

synergy. A coordination or cooperation between two or more agents or organs which results in smooth, economical activity.

synostosis. Restricted or lack of movement of a joint due to a bony connection.

synovia. A lubricating fluid secreted in joints, bursae, and tendon sheaths.

synovial. Pertaining to synovia.

systemic. Affecting the body as a whole.

tendon. A cord-like fibrous connective tissue serving for the attachment of muscles with points of origin and insertion.

thorax. That part of the body which is located between the clavicle and the diaphragm.

thyroid. The shield-like cartilage of the larynx which rests on the cricoid cartilage and furnishes an attachment for the vocal folds; also a gland.

tonus. A state of partial contraction of a muscle which produces a healthy, resilient quality in the muscle.

trachea. The cartilaginous and membranous tube extending from the larynx to the bronchial tubes; commonly referred to as the windpipe.

transverse. Usually applied to a plane which extends horizontally from one side of a structure to the other; a cross section.

tuberosity. Protuberance, eminence, or round process of a bone.

turbinate. Any structure shaped like a top and filled with pits, hollows or swirls; usually applied to bones of the nasal passage.

vein. A vessel which conveys blood toward the heart.

velum. The soft palate; the posterior and muscular portion of the roof of the mouth.

ventral. Toward the front or abdomen.

ventricle. A small cavity.

vertex. The topmost part; the top of the head.

vestibule. An antechamber; a small space or cavity at the beginning of a canal.

viscera. Generic term for the organs of any large body cavity; most frequently applied to the organs in the abdomen.

zygoma. The long arch that joins the zygomatic process of the temporal and malar bones.

CHAPTER SIX

APPLIED PHYSIOLOGY

An entire volume might now ensue, detailing the implications of the preceding chapters. Many facets must necessarily yield to limitations of space; nevertheless, some aspects of the *function* of the structures and entities described above demand elucidation.

Since tongue thrust appears in many guises, we think it advisable to have some system of classification by which to identify its various forms. Additionally, from an orofacial myological point of view, we would like to explore normal and abnormal function as applied to three specific components of the body: the tooth, the temporomandibular joint, and the nasal airway. Finally, we will offer some thoughts concerning the influences of musical instruments in this context.

Classification of Types

We have implied what must be obvious to anyone who has closely observed a sizable sample of tongue thrusters; there seem to be an infinite number of ways to swallow wrong. And it is true that certain individual peculiarities may be present that will not be seen again through a long run of cases. Nevertheless, a few patterns may be consistently seen; having identified the patterns, we may learn to predict oral behavior even before it is demonstrated to us. We are then ready to assay some basic classification of cases. We may actually understand the scope of the problem more completely once the broad outline of a classification system is before us. We become more aware of specific items and proceed with more assurance when we are not jolted by the appearance of a supposedly extraneous factor. We may also avoid some conflict and confusion when we finally leave off treating this problem as a single unified phenomenon and think instead of separate forms that may or may not have common characteristics. Moreover, there are important considerations for therapy in making some distinction between types.

165

Proposed Classifications

The decades of the 1950's and 1960's found several individuals concerned with categorizing the various manifestations that they observed. Several published their formulations, each based on a description of the resulting oral deformity and in some cases on the implied muscular malfunction that would suffice to create the diverse malformations.

Today, it is still not possible to trace each type back to its root cause, to note the peculiar circumstances that lead to this pattern of function rather than another, and thereby to offer a design based on etiology. Thus, we continue to label according to effect rather than cause.

The categories used by Barrett describe only those patterns of deglutition which result in overt harm to the dental structure. Some cases are presented in which there is reduced molar occlusion during swallowing, or a tongue posture that is lower than normal or even an observable thrust of the tongue during swallowing, and yet no associated dental impairment is apparent. It has been proposed that a category be provided for a benign, or nondeforming, deviant swallow. Such cases might instead be considered as a gradation of normal.

Tongue thrust is at root an orthodontic problem; it seemed logical, then, to base its classification as nearly as possible on the Angle system, the most widely accepted classification of malocclusion. Although it could not be carried beyond the first three categories, it nevertheless provides a point of departure, as may be seen in Table 6-1. Only major headings are listed; frequently recurring subtypes are usually found in each basic category.

Note that abnormal swallowing may be superimposed on any type of occlusion or may be a major etiological factor, depending on the individual case. Although certain patterns of behavior are characteristic of each type, the categories are so closely adjacent that some overlapping occurs. It is best to view this system as a continuous spectrum, with slight differences in behavior sometimes leading to one or another manifestation.

Each type appears to have its own set of characteristics extending far beyond mere differences in tongue action. These include classification of molar occlusion, incisal relationship, status of teeth during swallowing (apart or closed), the presence or absence of facial movement or contraction, and perhaps others. Much remains to be learned. We will pause here only to touch upon the tongue action displayed by each type.

Table 6-1. Classification of Types of Tongue Thrust

			TYPE
Anterior thrust	Anteroposterior	discrepancy	1. Incisor thrust (Angle Class I)
			2. Full thrust (Angle Class II, Division 1)
			3. Mandibular thrust (Angle Class III)
			4. Bimaxillary thrust
Lateral thrust	Vertical	discrepancy	5. Open bite
			6. Closed bite
			7. Unilateral thrust
			8. Bilateral thrust

Type 1. Pressure is concentrated on the incisors in a wedging action, driving uppers and lowers apart anteroposteriorally.

Type 2. "Dispersing" action of the tongue, spread between teeth around the dental arch from approximately first molar to first molar.

Type 3. Apex thrust against lower incisors or symphysis of mandible. Results in so-called "functional" Class III, a mandible of normal size displaced anteriorally.

Type 4. Tongue thrust against lingual margins of upper and lower incisal edges. May result in spacing of lower teeth, a condition that is quite unusual otherwise.

Type 5. Thrust into contact with lower lip before molars occlude.

Type 6. Flaccid generalized protrusion. Tongue usually engulfs the entire lower arch.

Type 7. Thrust at an angle toward the involved cuspid or bicuspid, somewhat like a misguided Type 5.

Type 8. Spread bilaterally between buccal teeth. Tongue tip usually braced against lower incisors in order to execute thrust, may thereby lead to Class III occlusion.

It is interesting to observe the relative incidence of the various types, although this, too, must remain somewhat inexact for the moment. The results of surveying 1000 consecutive cases in Barrett's practice yielded the results shown in Table 6-2. It should be noted that these cases were simply those referred for therapy by dentists and do not necessary reflect the relative incidence in the general population. For example, Type 1 may be more common than Table 6-2 would indicate, for since it is less deforming, it may be accepted by some persons without consulting a dentist. Again, some informal observations lead one to surmise that many Type 8 and some Type 7 patterns are not diagnosed even by dentists who are acutely aware of anterior thrust, and thus may also occur more frequently than indicated here.

Table 6-2.
Relative Incidence of Types Among 1000 Tongue Thrusters

Swallowing Pattern		Number	Percent
Type 1.	Incisal thrust	217	21.7
Type 2.	Full thrust	521	52.1
Type 3.	Mandibular thrust	18	1.8
Type 4.	Bimaxillary thrust	46	4.6
Type 5.	Open bite	89	8.9
Type 6.	Closed bite	57	5.7
Type 7.	Unilateral thrust	13	1.3
Type 8.	Bilateral thrust	39	3.9

Some of our historic differences of opinion should not be unexpected when one views a classification system such as that above, particularly those aroused by the presentation of a simplistic tongue thrust "syndrome," a term that does appear, at long last, to be on the wane. Of the five elements listed above, it should be noted that not one characteristic remains constant throughout the various types.

Forces Affecting the Tooth

It is our position that myofunctional disorders interfere with normal dental development and thereafter interfere with the orderly correction of dental anomalies. Some abnormal forces are so pervasive that they have repercussions throughout dentistry and myotherapy. These general saboteurs, individually or in coalition, may also provide the basis for many of the delimited intruders that plague us in only one special area. The sources of their power will be examined.

Tongue thrust is at the root of a problem of *pressure.* Nevertheless, it is not some independent agency capable of generating its own power, or a force that is foreign to the human mouth. Instead, it is a perversion of natural forces, a rearrangement of normal pressures into a harmful distortion. It might be enlightening, therefore, to examine some of the *normal* forces of the orofacial complex, noting as we go the consequences of disrupting or modifying each component of force.

A look at Figure 6-1 will give some idea of the stressful environment of the tooth. The illustration does not, however, reveal all the pressures to which the tooth is subjected, for this would require a third and even a fourth dimension, since time is also a factor. Thus, although the complete dynamics involved will demand some mental imagery, we may nonetheless use this figure as a point of reference.

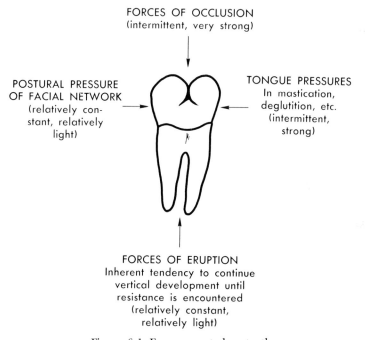

FORCES OF OCCLUSION
(intermittent, very strong)

POSTURAL PRESSURE
OF FACIAL NETWORK
(relatively con-
stant, relatively
light)

TONGUE PRESSURES
In mastication,
deglutition, etc.
(intermittent,
strong)

FORCES OF ERUPTION
Inherent tendency to continue
vertical development until
resistance is encountered
(relatively constant,
relatively light)

Figure 6-1. Forces exerted on teeth.

Forces of Occlusion

The forces of occlusion have been described in earlier chapters. They are powerful, occur only during normal mastication and deglutition, and are greatly reduced in abnormal function. The extent of the pressure parallel to the long axis of the tooth has been variously reported as from 100 to far more than 200 pounds per square inch in forceful occlusion. This force is essential and healthful, maintaining vitality in the mouth, strengthening the antigravity musculature, encouraging upright teeth in a symmetrical arch, and stimulating the deposition of firm alveolar bone of a dense texture.

Tongue thrust abates this critical force as the tongue is interposed between the dental arches. Antigravity muscles sag, fostering an undesirable orofacial rest posture. Molars are allowed to tip or wander. Once the tooth is malposed, any instance of occlusion then becomes an additional microtrauma, a force for evil rather than a beneficial influence.

Some degeneration of alveolar bone must also result. The so-called "law of transformation of bone" was formulated over a hundred years ago but has been confirmed in more recent studies. *Osteoporosis* indicates a reduction in the density of bone and may stem from lack of function or pressure. This process becomes more evident in extraction, when a tooth is left without an antagonist in the opposing arch. Alveolar bone surrounding such a tooth is easily penetrated by x rays and becomes a fading image on radiographs. It has been found, however, that restoration of pressure by the artificial replacement of an antagonist will cause rejuvenation of the alveolar process.

Force of the Tongue

The forces of the tongue are at work primarily during mastication and deglutition and may be reduced or absent during normal rest, speech, and other auxiliary functions. The combined strength of the eight muscles comprising the tongue represents a potential of some magnitude.

In normal function these forces are primarily directed against the palate, which is arched and otherwise so constructed as to withstand such pressures with ease; in fact, it depends on these pressures for its normal development. Any residual tongue pressure exerted in the horizontal plane during swallowing is readily absorbed by the teeth in normal occlusion: if the cusps are properly interlocked, the inclined planes provide reciprocal bracing and thus a stable, rigid wall that is self-sufficient; pressure is transmitted by this wall to the alveolar process, where it has a beneficial expanding effect.

So much attention has been accorded the tongue as an instrument of villainy—the creator of a variety of malocclusions in abnormal movement—that little comment has been made concerning the influence of the tongue in forming *normal* structures during dental development. Thus it might be advantageous to inspect this function for a moment. The tongue probably has little effect on the growth or shape of the basal bone; rather, the jaws develop toward a potential shape largely determined by heredity. This growth potential may be modified by poor nutrition or illness during a critical period or impeded by endocrine imbalance, ankylosis of the temporomandibular joint, trauma, etc. Nevertheless, basal bone is little influenced by the teeth it bears or the forces of its environment.

However, the normal structure of the alveolar processes, and the arrangement of the crowns of the teeth into the dental arch, are subject not only to the genes but also to other natural forces operating in the area. Strictly speaking, these structures—the teeth and the alveolar processes that support them—are *not skeletal features:* they grow *from* the skeleton but are nevertheless separate from it. They are parts of what is called the "exoskeleton" and are thus not subject to the same laws of growth and development that govern basal features. The shape of the initial arch is set by basal bone, but as the crowns of the teeth erupt away from skeletal bone into the oral environment, muscular forces supercede the diminishing influence of basal bone in the development and maintenance of the individual arch form.

The alveolar process is usually seen as a bone of convenience, or a bone of adaptation, the function of which is to support the teeth in whatever position they may assume. As such, the alveolar process does not possess the property of expansive growth and thus cannot widen the arch or effect tooth movement.

The tongue supplies the *only* sturdy, vigorous force on the dental arch from within outward and must play a major role in growth. Acting independently of the dental arch, this muscular organ is situated inside the denture and is ideally constructed to expand and maintain arch form. With glossectomy comes collapse of the dental arch.

The tongue is thus a vital element in normal development; it is from the essential nature of this required influence that the tongue derives much of its potential for evil. When its forces are displaced from palatal and alveolar locales directly onto the teeth by the tongue during deglutition, powerful abnormal forces are released with which the tooth is not prepared to cope.

Forces of Eruption

It has long been a basic tenet of dentistry that teeth seek occlusal antagonists: they tend to continue eruption until met by resistance.

However, the process does not stop even then; having made contact with an antagonist, the "prefunctional" phase of eruption comes to an end, but even so the process continues throughout the life of the tooth.

In normal function this moderate but constant force compensates for abrasion and other loss from the occlusal surfaces, maintaining the relative position of the teeth in the occlusal plane and thus their ability to meet in forceful relation. Removal of an antagonist, and failure to provide a replacement usually results in renewed eruption as the remaining tooth elongates in its search for resistance.

In many types of tongue thrust the interdental presence of the tongue unsettles the forces of occlusion and may result in the infraeruption of some teeth while allowing hypereruption of others. The teeth at the site of the thrust meet untimely resistance and stop erupting prematurely, leaving the familiar open bite. In the case of incisal thrust, the maxillary alveolar process itself may appear to be displaced superiorly. With retraction of the tongue in myotherapy, eruption resumes and the bite tends to close.

Quite a different effect is occasionally seen in teeth situated in a different part of the arch, away from the site of the thrust—that is, teeth that are held out of occlusion by tongue thrust but that do not contact the tongue. In some anterior thrusts, the molars may actually supererupt, thus opening the bite still more and requiring intrusion of the molars as part of orthodontic correction. In extreme cases, supereruption may even lead to mandibular resection.

Nevertheless, the force of eruption is a *normal* force, one that is routinely utilized to open the closed bite, as when a bite plate is inserted to hold molars out of occlusion until they erupt sufficiently to meet before incisors overclose. In a few cases we have found that some spontaneous intrusion of elongated molars has followed the establishment of firm molar occlusion in myotherapy.

Forces of the Facial Network

The muscles of the lips and cheeks are so interlaced and interconnected that action by one finds an echo in all; such reverberation is ordinarily mild. However, some of these muscles are so closely adapted to each other in their influence on dental structures as to function almost as a unit. One such unit has frequently been referred to as "the buccinator mechanism" (Fig. 6-2); it has also been called the "strap effect," since it can be seen that there is a continuous band of muscle fiber encircling the dental arches and anchored at the base of the occiput. Starting with the decussating fibers of the oribucularis oris, joining right and left fibers in the lips, upper and lower strands intermingle and run laterally and posteriorly around the corner of the mouth, connecting with fibers of the buccinator, which in turn insert into the pterygomandibular raphe just behind the dental arch. The fibers of the buccinator mechanism here interweave with fibers of the superior constrictor and continue posteriorly and medially to anchor at the pharyngeal tubercle of the first cervical vertebra, the atlas.

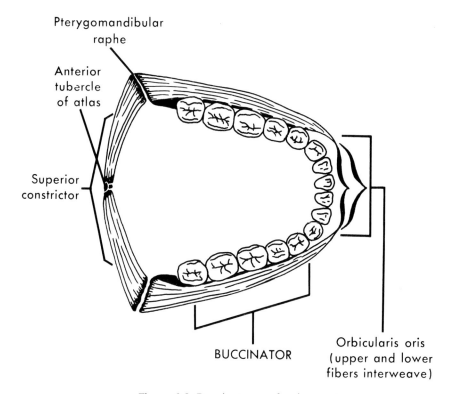

Figure 6-2. Buccinator mechanism.

In normal function, reciprocity among components is found, with adjustments in muscle activity available as compensations are required. It has been noted that even in the newborn the perioral musculature displays a smoothness in function quite in contrast to the jerky, un-coordinated movement of more distal or caudal muscles.

One function of this elastic band is to maintain the integrity of the dental arch by exerting a relatively constant enclosing force on the teeth.

It serves as the antagonist mechanism for the tongue. Even in postural rest there is the restraining influence of the buccinator mechanism. Although these forces have been found to be much less than the pressure of the tongue during deglutition, they are operative over a much longer period of time.

In abnormal swallowing, the compensatory movements noted above may work to restrain the malocclusion but may also serve to *increase* the discrepancy in some cases. An open-mouth rest posture, for example, removes the entire influence of the muscle band, allowing all anterior teeth to develop in labioversion, whereas an overdeveloped mentalis muscle may be accompanied by such constant and extreme force against the mandibular dentition that the lower incisors develop in linguoversion, or the contiguous skeletal jaw may be retarded in development. In the latter instance, even basal bone is influenced by soft tissue.

Forces of Torque

Although not a factor if each tooth is in proper occlusion, there are also rotating pressures—the twisting forces that are supplied by firm inter-cuspation of the teeth and to a lesser degree by the surrounding muscles. It is not uncommon for teeth to erupt at a slight tangent or in some posture less than optimal in relationship to adjacent or antagonist teeth. The normal forces of torque tend to correct such errors, uprighting and turning the teeth, and guiding them into a position of easy function as the inclined planes of posterior teeth slide down each other in forceful occlusion.

This force is supplied in adequate measure *only* during normal deglutition, since there is no other constantly recurring situation in which the cusps are brought into contact with force—not during mastication as we have seen earlier. These forces are thus largely unavailable to the teeth in abnormal swallowing, so that the cusps strike in an unhealthy manner during any remaining instances of tooth-to-tooth contact.

Forces of Mesial Drift

The forces of mesial drift fill in the last gap on the peripheral surfaces of the tooth and consists of distomesial pressure supplied by adjacent teeth at the contact points of the crown. This pressure was recognized by Angle, who referred to it as the "anterior component of force." It is the product of several factors, such as the cumulative effect of all orofacial muscle pressures and the axial inclination of the root structure, which is such that, when the teeth are brought firmly together, the crowns are propelled toward the front of the mouth, etc. It has the effect of closing gaps, which unifies and stabilizes the dental arch, and it compensates for wear at the contacting surfaces of adjacent teeth. It depends in part on continuity of the dental arch.

We noted earlier that mesial drift must be a concern for the pedodontist when early tooth loss requires a space maintainer to assure living space for a succedaneous tooth. To the contrary, this movement can be of assistance to the orthodontist when extractions are necessary during treatment, helping to restore and maintain contact of proximal teeth.

Some types of abnormal deglutition tend to disrupt the beneficial forces of mesial drift, as when mandibular or bimaxillary thrusts create diastemata in the lower arch. Almost any type of thrust poses a threat to the orthodontist in extraction cases, and even in some nonextraction instances: mesial drift may not be able to compensate for abnormal muscle pressure, thus destroying the integrity of the dental arch and resulting in unsightly spacing between the teeth after retention. Such unpleasant outcomes, discouraging for both dentist and patient, are precluded when myofunctional therapy is planned into treatment.

Atmospheric Force

Several authorities have assigned some value to atmospheric pressure, particularly during deglutition. It is postulated that, with a partial vacuum created in the oral cavity, atmospheric pressure in the nasal cavity would have the effect of forcing the bony palate to descent, especially during periods of rapid body growth and in the early years when the palatal bones are thin, are lightly calcified, and react readily to pressure. Descent of the palate helps to expand the upper arch, a highly coveted situation. Whether the force thus generated is genuine and sufficient has not been established. The vacuum itself would seem to provide a greater influence than the air driving downward from above the palate. In any event, such force as might be present would be totally lacking in deviate swallowing, with its reversal of intraoral air pressure.

Summary

With the possible exception of atmospheric force, it can be seen that the forces exerted on the tooth are so varied and of such magnitude that the tooth would surely be crushed and destroyed before long if nature had not had the foresight to endow the tooth with its hardy enamel armor and resilient periodontal ligament. Yet all these forces are *natural* forces, and *each force is essential* to the total health of the tooth. It is only when pressures are removed, disrupted, or displaced that trouble arises.

This may also hold true in the disordered mouth: a malocclusion represents nature's best attempt to maintain a balance among all the elements of the stomatognathic system. The clinician should be aware of a concept of which orthodontists frequently remind each other: an untreated malocclusion *is in dynamic balance at any particular time.* Disturbing this balance cannot be done with impunity; the dentition may be harmed by careless handling. If we keep in mind the relatively slight pressures employed in orthodontics to effect major physiologic response in the alveolar bone, we gain a proper respect for the state of equilibrium that may be only precariously maintained in a given mouth. Anyone proposing to tamper with this balance should have a very clear understanding of the dynamics involved and be prepared to compensate for every change, so that the end product is a revised equilibrium, not a still more harmful imbalance.

Temporomandibular Joint

As in the preceding section we must examine the *normal* physiology of the temporomandibular joint before we can appreciate its dysfunction or abuse. The musculature of the temporomandibular joint was discussed in the previous chapter. It is of interest that the structure of the joint, and the alignment of the muscles, provide a versatility of movement unique to the human species. A significant amount of controversy persists concerning the various positions and movements of the mandible and the implications arising from these; much of the strife regarding "centric" occlusion has its origin here. As was the case in Chapter Three, some care is required in traversing this area.

There is not even harmony with respect to the resting position from which movement starts. Some insist that normal posture should find the head of the condyle seated well within the depths of the fossa; others point out that the head of the condyle is angled slightly forward, and should hold the disc against the posterior slope of the articular eminence at rest.

There is agreement that we are not dealing with a simple hinge-like movement such as found in even the highest order of apes. The division of the joint by the meniscus into upper and lower compartments results in a compound joint. The statement of some that it is really two joints is hotly disputed by others, who feel that all movement is a product of combined action by both compartments.

Starting from a resting posture somewhat as seen in Figure 6-3, A, minor opening and closing movements probably occur between the cartilaginous disc and the head of the condyle, as in Figure 6-3, B. Opening the mouth widely, or protruding the jaw, creates movement between the disc and the fossa, the disc in this case moving with the condylar head, represented by Figure 6-3, C. Grinding movements elicit both types of action in the joint, the disc gliding forward and backward over the articular eminence while some rotation occurs between the disc and the head of the condyle. Figure 6-3, D, illustrates the result of extreme opening of the mandible.

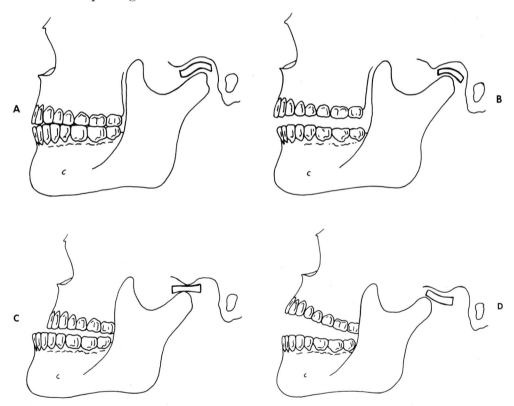

Figure 6-3. Function of temporomandibular joint. A, Normal resting relationship. B, Mouth slightly open, movement occurs between head of condyle and meniscus. C, Protrusion; meniscus moves with condyle. D, Dislocation; premature contraction of elevator muscles has locked head of condyle anterior to articular eminence, stretching all attachments of disc.

A recapitulation of muscle action might now be helpful. The mandible is *lowered* through contraction of the external (lateral) pterygoid, the anterior belly of the digastric, and the platysma muscles (minor influence may be supplied by the mylohyoid and geniohyoid). The mandible is *raised* by contraction of the masseter, temporalis, and internal (medial) pterygoid muscles. It is *protruded* by the simultaneous contraction of the external pterygoids on both sides. It is *retracted* by the contraction of the posterior fibers of the temporalis. *Lateral* movement is accomplished by contraction of the external pterygoid on one side only, and *grinding* movements represent a complex synergy involving many muscles, but characterized by the external pterygoids contracting in alternate fashion.

Note that the activities of this group of muscles in maintaining the position of the mandible are directly influenced by postural reflexes deriving from head and body position. Being subject to such modification, they are also easily stimulated by factors such as oral habits (mouth breathing, digit sucking, tongue thrusting, tooth clenching), emotional tension, respiratory movements and afferent impulses from oral structures.

We might also pause to note the complexity of the nerve supply to this area. It is served primarily by the many branches and fasciculi of four major nerves, the trigeminal (cranial V), the facial (cranial VII), the vagus (cranial X), and the great auricular, which is composed of cervical nerves 2 and 3. Fibers from these nerves anastomose (unite) with each other, with other branches of the same nerve, and with nuclei of the glossopharyngeal (cranial IX). The end product is a network of such intricacy that the brain is not always able to recognize the source of some neural excitations. This accounts for the phenomenon of "referred pain," when pain in a lower molar region often radiates to the TMJ. Pain in the joint may be interpreted as an earache or vice versa, and the content of the preceding paragraph is explicated.

Disorders of the Joint

The temporomandibular joint is subject to a wide variety of clinical problems. There are many reasons for its susceptibility. Among the more basic factors are the structure of the joint itself, the extremes of use and force to which it is subjected, its vulnerability to influences arising from tooth position, malocclusion, and other agents listed above. In addition, the neural innervation just described involves numerous small nerve centers located around the exterior of the skull, particularly in the region just superior to the TMJ.

Infection in the joint is rare. Tumors occasionally develop in various components of the joint, creating painful excessive pressure on surrounding nerves. Certain types of blows can snap the neck of the condyle, requiring splinting of the mandible until the fracture is reduced. None of these concern the oral myologist. However, almost all of the remaining disorders incorporate aspects of muscle dysfunction, either as a causative factor or as a resultant anomaly.

Dislocation. When the mandible is lowered excessively, opening the mouth to an extreme degree, the head of the condyle may slip out of the fossa entirely and come to rest in front of the articular eminence, resulting in dislocation. Because of its looseness, the capsule does not tear in this situation, but is stretched to a harmful extent, as are the muscles and ligaments of the TMJ.

In normal closure, the precise sequence of muscle contraction begins with the posterior belly of the temporalis, which retracts the condyle, bringing the head into position behind the peak of the eminentia. The elevators then replace the head into the fossa. If this sequence is disturbed, and the elevators contract before the mandible is retracted, the jaw will move superiorly while still beneath or slightly anterior to the articular eminence, and the mandible is dislocated (see Fig. 6-4, D).

The resultant pain is usually severe, causing muscles to go into spasm; continued spasm of the elevator muscles maintains the punishing malposition, and outside assistance is required in most cases to reduce the dislocation. When such help is needed, or when the dislocation is complete as described, it is called *luxation.* When the dislocation is partial, or when the patient is able to retract the jaw himself/herself, the term is *subluxation.*

Some patients have stretched the capsule and the supporting ligaments to such an extent that they experience constant subluxation. This has several causes, but the effects are invariably harmful. Mechanical influences such as malocclusion should be eliminated first, after which the orofacial myologist may be helpful in toning the antigravity muscles and in disrupting the habitual thrusting of the jaw which may continue after mechanical correction. A mandibular tongue thrust will frequently be found in these cases.

Two terms, *clicking* and *crepitus,* should be mentioned in this context. They are often used interchangeably, but have slightly different connotations.

Clicking. During some excursions of the mandible, many patients hear or feel a click or snap, a sharp popping sensation. This is the sound resulting from the meniscus snapping back and forth over the articular eminence in excessive fashion. This condition has several causes, such as reduced thickness of the disc; or the fibrous attachments of the disc to the condyle may become stretched, causing hypermobility of the meniscus; or constant overstretching of the capsule may result in loss of resiliency, a "sloppy" joint that is quite conducive to clicking.

Crepitus. As a general rule, a condition somewhat more severe is indicated by crepitation. Crepitus is usually defined as the sound emitted when the broken ends of a bone are rubbed together. It has also been likened to the sound of walking in loose gravel. It is often the result of a perforated disc; in this case it is the sound of bone-to-bone contact between the head of the condyle and the fossa or eminentia.

Arthritis. The temporomandibular joint may be the site of arthritis, occasionally even in younger patients. Arthritis is defined as inflammation of a joint. There are two general type of arthritis, that of infectious origin and that of noninfectious origin. We have noted that the TMJ is rarely subject to infection, although in a few cases it is invaded as an extension of generalized arthritis in all joints. More commonly it is prey to the noninfectious type, specifically *osteoarthritis.* Osteoarthritis is characterized by destruction of the articular fibrocartilage, overgrowth of bone, especially around the rim of the fossa, spur formation on the condylar head, and thus impaired function. Osteoarthritis is usually the result of malfunction of the masticatory mechanism, growing out of malocclusion or other harmful jaw relationship, or of improper function of the muscles primarily involving hypercontraction. Any circumstance that brings undue stress or pressure to bear on the protective disc can reduce the cushioning effect of that cartilage and lead to inflammation.

Intolerable stress is avoided by some patients through the use of an unconscious procedure of which the oral myologist should be aware. This is the "myoprotected occlusion" delineated by Picard (1975). After outlining a "mutually-protected occlusion" in which all elements of the stomatognathic system work in harmony to prevent trauma in any one segment, he provides a contrast. He itemizes the steps leading from an occlusal discrepancy, through self-equilibration (bruxing) with the resultant additional stress damaging the investing tissues, to pulpitis, periodontal breakdown, TMJ dysfunction, occlusal disease, and pain. When the system exceeds its limit of tolerance, the tongue may be placed as an interocclusal mass during swallowing, since it is then that the teeth occlude with force. Closing on the padded cushion of the tongue relieves all previous ills, and a myoprotected occlusion is instituted. Picard warns that attempts to alter deglutition before the original causative factors are corrected would deprive the system of its last protection.

Other Degenerative Conditions. In contrast to the arthritic tendency toward formation of extraneous bone are those conditions leading to degeneration and loss of osseous tissue. It is not uncommon to find patients with the severe protrusion accompanying some Class II, Division I malocclusions who habitually carry the mandible forward. This positions the head of the condyle against the descending slope of the eminence with some force; considerable tension in the external pterygoids is required to maintain such a posture. The result may be bone resorption in both the head and the eminence at the focal point of the abnormal pressure. This may also be true of many functional Class III patients; in this case, the tongue thrust jams the mandible forward, creating the injurious condyle-fossa relationship.

Something similar is found in patients with an open bite, but for a different reason. In normal function, protrusion of the mandible causes the lower incisal edges to slide down the lingual surface of upper incisors, instantly disarticulating the molars. This prevents the bumping and scraping of cusps between upper and lower molars. Such incisal guidance is lost to the open bite patients, who must then rely on an abnormal function of the joint itself to separate the posterior teeth, a process that may prove traumatic to the TMJ with constant repetition over a period of years. Failure to disengage the molars before protrusion occurs is equally harmful, with repercussions throughout the stomatognathic system; it has the same effect on the neuromuscular system as a "premature contact" or other occlusal interference. Premature contacts may trigger muscle spasm, and the vicious circle of consequences is thereby reinforced.

According to several authorities, a majority of TMJ problems arise from insufficient vertical dimension. In younger patients this may result from delayed eruption of molars, while in adults it commonly follows the extraction of molars without providing replacements. In these cases, the head of the condyle is forced upward and somewhat forward; laminographic x-rays frequently reveal some erosion of the fossa and a decidedly "moth-eaten" anterior surface of the head of the condyle.

Myofascial Pain-Dysfunction Syndrome. Recent years have seen an outpouring of literature dealing with this subject, but referring to it under five or six titles; temporomandibular joint myofascial pain-dysfunction syndrome, though lengthy, is one of more common labels. Calling it a "TMJ syndrome" is less precise, but saves time.

The incidence of this condition in the general population has not been established, but it appears to be very widespread; estimates exceeding 20 percent have been made. The symptoms, all stemming from dysfunction of the TMJ, include pain of the head, face, ear, neck, and shoulder, in various combinations; intermittent vertigo; tinnitus (subjective ringing ear noise); burning sensations in the sides of the tongue; spontaneous subluxation; clicking and crepitation of the joint; occasionally, trismus or restricted jaw movement; difficulty in swallowing; fatigue; forgetfulness; and muscle spasms throughout the body.

The pain-dysfunction syndrome may lead to arthritis of the TMJ as a secondary effect, but is a distinct condition resulting primarily from dysfunction of the masticatory muscles. A majority of patients suffering from this condition never receive proper treatment. They tend to consult physicians who lack dental expertise. They are placed on migraine programs and medications, which prove futile. Their vertigo is ascribed to untreatable inner ear conditions. They are classified as neurotic and referred for psychotherapy. It is true that they may require the combined services of physician, dentist, psychologist, and oral myologist before they are restored to full health; nevertheless, a knowledgeable dentist can remove most of their overt symptoms with surprising speed, usually within one or two months.

A point to be stressed is that the orofacial myologist can perform a great service by being aware of the pain-dysfunction syndrome, particularly during initial examinations. It is helpful to check such items as: pain upon palpation of masseter and temporalis muscles; clicking; limited opening; lateral deviation of the mandible upon opening; tinnitus; frequent headaches; and evidence of bruxing. Inquiry into these areas usually reveals the presence and extent of any current problem, and may even unmask a potential hazard. Treatment may then be instituted as indicated, or referrals of an appropriate nature can be made.

Summary

The treatment of problems arising from the temporomandibular joint, as with other orofacial myofunctional disorders, remains the scene of confusion and controversy. It is helpful to realize that almost every affliction of the TMJ is basically a result of abnormal muscle function, and almost always stemming from tension and overcontraction of the muscles. In normal function, the teeth are in firm occlusion only during the act of deglutition, and regardless of the estimate of swallowing frequency, the musculature should be in strong contraction a cumulative total of only a few minutes per day.

However, we live in a stressful society. A universal reaction to stress is sustained TMJ muscle contraction (clenching) and continual overworking of these muscles (grinding) with a resulting combined force far in excess of the limits for which this system was designed. When perpetuated over a period of years, some breakdown must be anticipated.

The well-grounded orofacial myologist should understand the composition and function of the temporomandibular joint, should be cognizant of some of the symptoms of its dysfunction, and should be prepared to treat the ensuing or causative muscle defects.

The Nasal Airway

The individual osseous parts of the nasal airway were laid out in the preceding chapter, but somewhat on the order of the scattered pieces of a puzzle. We may now assemble them into a working whole, drape them with membrane, and examine their function.

The Nasal Cavity

We deal here not with a mere bony skeleton, but with a nose completed by a lower framework of five major cartilages: two lateral, two alar, and the septal cartilage. Several other small pieces fill in odd-shaped crannies near the face. All are connected to each other and to the bones by tough fibrous connective tissue.

The entrance to the nasal cavity is through the *nares,* or nostrils, and the exit into the nasopharynx is through the two *choanae.* Immediately within the nostril is a dilation, the *vestibule,* that is lined with skin which in turn gives rise to stiff, coarse hairs, the *vibrissae.* The skin of the vestibule blends into the mucous membrane, which thereafter lines the passageway throughout.

The medial wall of the passage, the septum, is smooth and flat, the roof very narrow except posteriorly, the floor gently curved from side to side and almost horizontal anteroposteriorly.

The lateral wall is more interesting. This small area is composed of bits of six different bones: the lacrimal, ethmoid, sphenoid, palatine, and maxilla, in addition to the inferior nasal concha, which is a separate bone. The lateral wall is made irregular by the projection of three *conchae*, or *turbinates*. They are called conchae because of their shell-like shape, and turbinates because of their function. They curve gently upward front-to-back, and extend horizontally into the nasal cavity, almost reaching the septum. In effect, they thus subdivide each nasal cavity into three groove-like passageways, the *nasal meatuses*. Each meatus is named for the turbinate above it; the inferior meatus lies below the inferior concha, and the middle and superior meatuses are below the middle and superior conchae, respectively. The *nasolacrimal canal* opens into the inferior meatus; it drains tears from the eyes into the nose and accounts for a "runny nose" when crying.

The Paranasal Sinuses

Grouped about the nasal passage is a system of cavities, the paranasal sinuses, all of which are pneumatic areas and all of which connect with the nasal passage. While the sinuses serve a variety of purposes, to count them off as mere resonating chambers for the voice, or as a means of lightening the weight of the skull, as is frequently done, is to underestimate their value; for the orofacial myologist, they have a much greater importance. In the well-ordered nasal passage they perform a wholesome and gratuitous service by expanding and facilitating the labyrinthine channel; in the unused nose they lurk as a morbid and hostile menace as they constrict and impede the airway.

These sinuses are located in four bones, the frontal, ethmoid, sphenoid, and maxilla. They are rudimentary in early life, only the maxillary sinus being definitely present at birth. Through the process of *pneumatization*, bone around invading mucosal sacs is resorbed and the cavities enlarge. The frontal and sphenoid sinuses are radiologically visible at six or seven years of age, while the ethmoid cells, small but numerous, do not develop adult proportions until puberty.

The frontal sinuses communicate with the nasal passage through a short canal, the *fronto-nasal duct*, which opens into the middle meatus. The sphenoid sinuses and the posterior ethmoid cells open into an aperture above the superior concha, the *sphinoethmoidal* recess; other ethmoidal cells open into the middle meatus. The maxillary sinuses, the largest of these cavities, occupy the body of the maxilla and communicate with the middle meatus by way of an opening in the upper medial wall of the sinus; since the sinuses lie just external to the lateral nasal wall, drainage is poor in the erect posture, requiring the laying of the head on one side.

The Mucous Membrane

Mucous membrane lines the entire nasal cavity except the vestibule. It is continuous throughout all of the chambers with which the nasal passage communicates, the nasopharynx behind and the paranasal sinuses above and laterally.

The epithelium forming the surface of the mucous membrane is ciliated; that is, the free ends of the epithelial cells give rise to cilia, microscopic hair-like processes. Eight cilia project from each cell, and beat in a wave-like motion at an incredible rate toward the pharynx; that is in distinction to the bronchi, where the cilia beat *upward*, waving mucus, pus, and dust particles out of the lungs but also toward the pharynx.

The mucous membrane is thick and richly endowed with blood vessels, especially over the conchae; it is thinner in the meatuses and in the sinuses. It contains a layer of glands throughout, in which both mucous and serous alveoli are present; it thus emits both a thick mucoid secretion and a thinner, serous fluid which contains both an acid and the bacteria-destroying substance, *lysozyme*, which is also present in tears. The combined secretions form a sheet which overlies the cilia rather than the epithelial cells proper.

Function of the Nasal Airway

We are now prepared to put the nasal airway into operation. It might be considered a processing plant for breath, warming, cleansing, humidifying, even sterilizing the air. As breath is inspired, the vibrissae guard the twin openings of the nares, removing gross particles of foreign matter. Since the nasal passage is really quite narrow, air is drawn over the nasal mucosa in a thin stream. The turbinates set the air in whirling motion, speeding its rate, forcing it into sinuses, and circulating it within the nasal fossae. The air is thus brought into intimate contact with the sticky mucous membrane, which traps and holds such microscopic matter as pollen, dust, and bacteria.

Body heat, radiating from venous cavities under the mucosa of the turbinates, effectively warms the inspired air. At the same time, secretions in the nose and in the paranasal sinuses combine in the humidifying process; it is estimated that approximately one quart of fluid is secreted each day in order to properly humidify the breath.

Through the combined actions of the cilia and respiration itself, the clogged and bacteria-laden mucoid blanket is removed and quickly renewed as it is moved posteriorly to be swallowed, blown away, or expectorated.

Mouth Breathing. In juxtaposition to the above, we should examine the process of mouth breathing. The antigravity muscles of the mandible and tongue are provided adequate development only during normal deglutition. When robbed of tonicity by tongue thrust, the sheer weight of the mandible, tongue, and their associated structures proves too great for many patients, and in a resting state the mouth falls open. Even though the nasal airway is available, and is used occasionally, the more common result is oral breathing.

All of the beneficial contributions of the nasal airway are now surrendered. Raw, dry, dirty, cold air is drawn into a respiratory system not prepared for such onslaught. Large quantities of moisture are evaporated from the oral and pharyngeal mucosa, much of it lost to the body during exhalation. The lower lip becomes parched and chapped as the air fans back and forth over its surface. The pharyngeal wall becomes less sensitive. The alveoli of the lungs are impaired, along with the ability of the pulmonary alveolar epithelium to perform its essential function in exchanging dissolved gases between the breath and the blood. The mucoid sheet in the nasal passage and sinus cavities, tops moving and stagnates; it thickens, which not only reduces the diameter of the passageway, but may also irritate the tissues beneath, causing edema and still further reduction of the airway. Eventually, the nose may become blocked, and the patient has an excuse for mouth breathing.

The nasopharynx may also pose a threat in this situation, particularly during sleep, when swallowing incidence drops dramatically. With saliva being evaporated by the breath stream, the need to swallow is reduced still further. Stagnating secretions in the nasopharynx may then follow an alternative route into the bronchial tree, given a supine position from which droplet infection into the lungs is most likely to occur.

At least some mild sinusitis is commonly observed in mouth breathing patients. Even this subverts the mucous linings of the sinuses from their normal function in respiration.

Respiratory Allergy. A number of patients present themselves for therapy who suffer from true respiratory allergies. When severe and constant, such allergies are felt to be a contraindication for therapy, and we tend to postpone such efforts until the condition is improved. Little achievement of lasting worth can be maintained against the deleterious effects of mouth breathing, and when the result of mouth closure is anoxia, progress in therapy is not a realistic aspiration.

Nevertheless, it has seemed that the milder allergies have proved no insurmountable barrier. In fact, our routine procedures have often seemed to be a boon to the patient who makes a sincere attempt. As s/he is provided with muscular ability for effortless mouth closure, and gradually encouraged in nasal breathing, the airway has appeared to improve and the allergic reaction lessen.

This clinical impression was reinforced by a study done by Toronto (1970), who examined a group of former patients of one of the writers (R.H.B.), not only with a view to determining the number who had relapsed, but if possible *why* they had failed. The one consistent finding in the relapse group was continued mouth breathing. However, this group did not include most of those who had originally reported respiratory allergy; instead, they were primarily those whose record revealed poor performances during therapy, who failed to return for follow-up procedures, and who had exerted themselves as little as possible. This led Toronto to the conclusion that while mouth breathing is a major cause of relapse to tongue thrusting, excessive allergies are not.

Summary

Oxygen is the most imperative requirement of the human body; food, liquid, and other needs can be postponed for some time, but life is brief without oxygen. It is therefore not difficult to understand the resistance of some patients to any suggestion that might imperil their source of supply for this precious commodity. Yet a few myotherapists reveal some exasperation when their assignments are not met with enthusiasm. Some patience is required. Without a reasonably healthy nasal airway, functioning in a reasonably normal manner, all is lost.

Effects of Playing Wind Instruments

When public school music departments tune up each fall, it is not uncommon for parents to seek advice about the myological and dental influences of certain musical instruments. The orofacial myologist should be able to make an enlightened response, since the choice can have an effect on therapy. The clinician's interest will focus mainly on lip and tongue postures; however, the conscientious therapist should also be aware of the mechanical force of the instrument itself upon the teeth. Moreover, the clinician should be prepared to explain the desirable qualities of a beneficial instrument, rather than merely condemn a harmful one.

As we examined the available literature on this subject, five significant articles were found in national dental journals. These were written by Strayer (1939), Pang (1976), Parker (1957), Herman (1974), and Wiesner and co-workers (1973). Two of them (Pang and Parker) report on research done by the writers, while the other three authors write from clinical experience or report of the work of others.

Pang's (1976) research design was a double-blind approach, wherein dental occlusions of 76 seventh-grade students were examined. It is difficult to determine the numbers of subjects in the control, experimental, and subgroups in this research, but a reasonably close calculation would be 46 in the experimental group and 30 control subjects. It is also not clearly stated whether all the experimentals had malocclusions, but members of the control group did all have malocclusions. There were roughly equal numbers of boys and girls. There were 19 Class A instrument players (trumpet, bugle, French horn, trombone, tuba); 21 Class B players (clarinet, saxophone); two Class C players (oboe, bassoon, English horn); and four Class D players (flute, piccolo).

Pang's subjects were examined orthodontically before and after six months of playing in school bands. Control subjects played no musical instruments. A summary of the results of his research is included in Table 6-3.

Parker's (1957) study was a roentgenographic analysis of the physiology involved in playing wind instruments. Subjects were 84 school pupils of all ages and both sexes. Excellent data were gathered showing how students position the instruments in the mouth, but several unwarranted conclusion, regarding the effects of playing instruments, were drawn from this research. Subjects were photographed only once, and no before-and-after data were taken. The control group consisted of 30 youngsters with Class II, Division I malocclusions who did not play a musical instrument. The average angle of the central incisors of this latter group was found to be 116.2 degrees, compared to a very consistent 107.6 to 109.0 degree means in the four subgroups of experimental subjects. Parker erroneously concluded that this demonstrated an absence of deleterious effects upon dentition from playing wind instruments. He wrote, "The study indicates that, if the patient is under the proper supervision and is using the correct embouchure, the orthodontist need not concern himself about a clarinet, flute, or saxophone player who may have a Class II, Division I malocclusion."

Table 6-3. Wind Instruments and Occlusion*

Type of Instrument	Strayer (theory)	Herman (theory)	Pang (research)
A. Trumpet, trombone, tuba, French horn	Good for overjet; bad for maxillary retrusion	Good for Class II, Division I (overjet)	Good for overjet
B. Clarinet, saxophone	Good for maxillary retrusion; bad for overjet	Good for Class I, Class III, bad for Class II (overjet) and for irregular, sharp teeth	Little effect on overjet, may cause open bite
C. Oboe, bassoon, English horn	Good for hypotonic or short lips; bad for "complicated" Class I's"	Good for open bite; good for Class II (overjet)	Good for overjet; may cause open bite
D. Flute, piccolo	Good for Class I and Class III, short upper lips, mentalis habit; bad for overjet	Good for Class II (overjet)	Bad for overjet

*A comparison of the conclusions of three writers regarding the effects of playing wind instruments on dental occlusion: Strayer (1939), Herman (1974), and Pang (1976).

The articles by Strayer (1939), Wiesner (1973), and Herman (1974) were based either on their own experiences with wind instrument players or reports in the literature. The Wiesner paper includes a chapter on the use of protective appliances recommended for musicians, to prevent injury to soft tissues from orthodontic appliances and sharp, irregular teeth.

Table 6-3 summarized claims and findings of Strayer (1939), Herman (1974), and Pang (1976). There is a general agreement among the three writers on the beneficial and harmful effects of playing certain instruments.

The trumpet and related instruments can reportedly affect overjets favorably. Subjects with overjet should avoid playing the clarinet, saxophone, flute, or piccolo. The oboe, bassoon, and English horn may affect an overjet favorably.

Since only one of the five writers mentioned in the first paragraph of this section actually studied cause and effect, perhaps the opinions of the others should be regarded with less credence. Nevertheless, conclusions reached from clinical experience need not be rejected as invalid, especially when they are in essential agreement with research findings.

As in the treatment of any aspect of human behavior, it is wise to consider the generalizations others have made as you examine the individual patient. Before concluding that the playing of a given instrument will help or harm the teeth or lips of a child, however, several variables should be considered.

1. Relationships between the dental arches: in order for a child with an extreme Class II or Class III occlusion to play any brass instrument he will be required to exert strong effort to align the upper and lower lips. Pang's (1976) warning about the possibility that playing a clarinet, saxophone, oboe, bassoon, or English horn may cause open bite should be taken seriously, especially when the patient has a tendency toward such a malocclusion.

2. Intra-arch conditions, such as sharp, irregular maxillary incisors, may result in injury to the lips.

3. Improper positioning of the instrument may produce a lever-like action against the upper incisors.

4. The mouthpiece may be forced against the upper alveolus, lingual to the incisors, and periodontal damage may result.

5. Hypotonia and shortness of the lips may be aided considerably by playing a brass instrument. A flute or piccolo may also assist in repositioning a short upper lip, as long as overjet is not pronounced.

6. It is possible that a child with an anterior overjet can be aided by playing a trumpet, but only if the tongue is properly held away from the lingual surfaces of the upper incisors.

7. The eventual necessity for a retainer should be considered. Some forms of retainers reduce the vertical dimensions of the maxillary arch to such a degree that it becomes very difficult to play certain wind instruments.

8. Instructors who are unfamiliar with relationships between the dentition and instrument-playing may teach an embouchure that is either

difficult for, or harmful to, the patient. Others who are knowledgeable may have 50 or 60 students in the school band, and hence not become aware of the dental problems of individual players.

Recommendations

Our practice is to consider each patient individually. Interarch and intraarch dental factors, as well as soft tissue conditions are examined carefully. If the child is in the process of selecting an instrument to play, there are several which definitely will not harm the teeth. If s/he is already playing one that is potentially harmful, s/he is asked to bring the instrument to therapy and demonstrate its positioning as s/he plays it. If there is any possibility of damage to the dentition or soft tissues, the clinician contacts the music teacher to explain the situation and ask whether another type of embouchure might be taught. If that is not feasible, it is recommended that s/he discontinue playing the instrument.

We are conservative in our estimates of improvement to be expected in the dentition or soft tissue from proper playing of an instrument. A child who practices regularly, for long periods of time, may improve the tonicity of the lips, however, and the playing of the instrument may be a helpful adjunct to therapy.

In this area of study, as in all others related to oral myofunctional problems, longitudinal research is needed to determine *long-term* effects on the players with and without malocclusions or adverse orofacial conditions.

REFERENCES

Blitzer, M. H. (1977, May). Diagnosing the TMJ pain dysfunction syndrome. *Prosthodontics, 8,* 37–38.

Bohl, C. F., and Knap, F. J. (1974, July). Evaluating occlusal relationship, mandibular dysfunction, and temporomandibular joint pain by palpation. *J. Prosthet. Dent., 32,* 80–85.

Carlsson, S. H., Gale, E. N., and Ohman, A. (1975, September). Treatment of temporomandibular joint syndrome with biofeedback training. *J. Am. Dent. Assoc.*

Costen, J. G. (1934, March). Syndrome of ear and sinus symptoms dependent upon disturbed function of the temporomandibular joint. *Ann. Otol. 43,* 1.

Emslie, R. D., Massler, M., and Zwemer, J. D. (1952). Mouth breathing: I. Etiology and effects (a review). *J. Am. Dent. Assoc., 44,* 506–521.

Francis, C. C., and Martin, A. H. (1975). *Introduction to human anatomy* (7th ed.). St. Louis: C. V. Mosby.

Fried, L. A. (1976). Anatomy of the head, neck, face, and jaws. Philadelphia: Lea and Febiger.

Gelb, H., and Tarte, J. (1975, December). A two-year clinical dental evaluation of 200 cases of chronic headache: the craniocervical-mandibular syndrome. *J. Am. Dent. Assoc., 91,* 1230–1236.

Graber, T. M. (1966). *Orthodontics: principles and practice* (2nd ed.). Philadelphia: W. B. Saunders.

Grant, J. C. B. (1972). *An atlas of anatomy* (6th ed.). Baltimore: Williams and Wilkins.

Greene, C. S., and Laskin, D. M. (1974, December). Long-term evaluation of conservative treatment for myofascial pain-dysfunction syndrome. *J. Am. Dent. Assoc., 89,* 1365–1368.

Herman, E. (1974, September). Dental consideration in the playing of musical instruments. *J. Am. Dent. Assoc., 89,* 611–619.

Leech, H. L. (1958). A clinical analysis of orofacial morphology and behavior of 500 patients attending an upper respiratory research clinic. *Dent. Parct., 9,* 57.

Overstake, C. P. (1975, July). Investigation of the efficacy of a treatment program for deviant swallowing and allied problems, Part I. *Int. J. Oral Myol., 1,* 87–104.

Overstake, C. P. (1976, January). Investigation of the efficacy of a treatment program for deviant swallowing and allied problems, Part II. *J. Oral Myol., 2,* 1–6.

Pang, A. (1976, March). Relation of musical wind instruments to malocclusion. *J. Am. Dent. Assoc., 92,* 565–570.

Parker, J. H. (1957, June). The Alameda instrumentalist study. *Am. J. Orthod., 63,* 399–415.

Picard, P. J. (1975, January). Gnathology and the myoprotected occlusion, a hypothesis. *Int. J. Oral Myol., 1,* 78–82.

Pomp, A. M. (1974, September). Psychotherapy for the myofascial pain-dysfunction syndrome: a study of factors coinciding with symptom remission. *J. Am. Dent. Associ., 89,* 629–632.

Schwartz, L., and Chayes, C. M. (Eds.) (1968). *Facial pain and mandibular dysfunction.* Philadelphia: W. B. Saunders.

Shore, N. A. (1959). *Occlusal equilibration and temporomandibular joint dysfunction.* Philadelphia: J. B. Lippincott.

Sicher, H., and DuBrul, E. L. (1975). *Oral anatomy,* (6th ed.). St. Louis: C. V. Mosby.

Stack, B. C., and Funt, L. A. (1977, January). TMJ dysfunction from a myofunctional perspective. *Int. J. Oral Myol., 3,* 11–26.

Stone, S., Dunn, M. J., and Robinov, K. R. (1971, Fall). The general practitioner and the temporomandibular joint pain-dysfunction syndrome. *J. Mass. Dent. Soc., 20,* 262–268.

Strayer, E. F. (1939, April). Musical instruments as an aid to the treatment of muscle defects and perversions. *Angle Orthod., 9,* 18.

Swinehart, D. R. (1950). The importance of the tongue in the development of normal occlusion. *Am. J. Orthod., 36,* 813–830.

Toronto, A. (1970). *Permanent changes in swallowing habit as a result of tongue-thrust therapy prescribed by R. H. Barrett.* Unpublished thesis, University of Utah, Salt Lake City, Utah.

Weinberg, L. A. (1974, September). Temporomandibular dysfunction profile: a patient-oriented approach. *J. Prosthet. Dent., 32,* 312–325.

Wiesner, G. R., Balback, D. R., and Wilson, M. A. (1973). *Orthodontics and wind instrument performance.* Music Educators National Conference, Washington, D.C.

Woodburne, R. T. (1973). *Essentials of human anatomy,* (5th ed.). New York: Oxford University Press.

CHAPTER 7

ETIOLOGIES

Etiologies

Tongue thrust is a normal behavior during infancy. A search for causes, therefore, must focus on factors that contribute to the *persistence* of tongue thrust, rather than those that might be responsible for its original occurrence. The list of etiologies that has been proposed over the years is a long one. It includes bottle feeding, environmental pressures, neurotic tendencies, malocclusions, sucking habits, mouth breathing, allergies, genetic structural characteristics, open spaces during mixed dentition, enlarged tonsils and adenoids, and prolonged soft diet. Any of these may be contributing factors in the persistence of tongue thrust in a given patient.

1. **Bottle Feeding.** In the past, much credence was given to the theory that improper feeding of infants resulted in tongue thrust. Straub (1960) had reported that of 478 tongue thrusters he had personally seen, only two had been breast-fed, and that the mothers of these two infants had had a tremendous supply of milk that flowed freely. He attributed the abnormal swallow to a too rapid flow of milk that caused the infant to thrust the tongue forward in order to prevent being choked. The nipples used on most baby bottles were so long they almost reached the back of the throat, making it extremely difficult for the infant to place the tongue against the roof of the mouth. The problem was aggravated by the practice of mothers who placed extra holes in the end of the nipple, or enlarged the ones that were already there, to make sure that the milk would flow rapidly. The resulting thrusting action of the tongue supposedly also occurred in infants whose mothers had rich supplies of milk.

Picard (1959) explains that in natural feeding the orbicularis oris acts as a seal against the breast; however, mere negative pressure does not suffice to deliver milk into the infant's mouth. A pumping action of the mandible, allowing direct lingual pressure against the nipple to alternate with suction, is required to extract the milk. The milking action thus requires a protrusion of the mandible out of its infantile distal position and the expenditure of considerable effort. It has been found that the digastric muscle is approximately twice as strong in the newborn as it is in the adult.

In bottle feeding, Picard continues, the bottle is held in such fashion and the hole in the nipple kept at such diameter that the easy flow of milk through light sucking alone is ensured. Infants receive milk in such quantities that they must learn to cope with a surplus rather than to work their jaws for their daily meals. Instead of receiving nutrition as a reward for hard work, they receive it without effort, which starts them out just right for our push-button civilization.

During the last two decades, considerable evidence has been found that questions the relationship of breast feeding to tongue thrust. Subtelny (1965) cites several studies of breast feeding that indicate that differences between swallowing habits in breast-fed babies and those in bottle-fed babies are minor. Weinberg's (1970) review of research revealed no significant relationships between manner of feeding and existence of tongue thrust in children. He listed the following findings of research:

1. The tongue tip protrudes well beyond the mandibular ridge in newborns during breast feeding and during bottle feeding. Patterns of movement in the two types of feeding are not significantly different (Adran and Kemp, 1955).

2. No significant difference was found between duration of breast feeding in infancy and distal occlusion of the mandibular teeth. Subjects were 1085 preschool children (Bijlstra, 1958).

3. No significant relationship was found between manner of feeding and maxillary protrusion in 1000 children ages 6 to 12 years (Bijlstra, 1958).

4. There is no greater prevalence of tongue thrust in bottle-fed children (Leech, 1958).

Hanson and Cohen (1973) studied 214 children, 4½ to 5 years old, and found approximately 50 percent to be tongue thrusters. There was no significant difference between the number of tongue thrusters who had been breast-fed and the number of normal swallowers who were breast-fed. When the children were seen again at the age of 8 years, no significant relationship was found between bottle and breast feeding and the retention or development of the tongue-thrust swallow.

Despite convincing evidence that no relationship exists in large groups of subjects between manner of feeding in infancy and persistence of tongue thrust, such a relationship may occur in a given child.

2. **Genetic Influences.** Among patients we have seen, often the face of the child, together with lingual habit patterns, mirror those of the mother and father. Patients have reported that their orthodontist has said something like "Your child has inherited his mother's lower jaw and his father's upper jaw." Of course, it is not that simple. We do not inherit one parent's entire set of mandibular characteristics and another parent's complete maxilla. There is a complexity of factors, numberless combinations of which might predispose a child toward a tongue thrust pattern. Among them are:

(a) a tendency toward allergies and upper respiratory congestion;
(b) an extremely high or narrow palatal arch;
(c) an unusually large tongue;
(d) a restricted nasal passageway due to small nares or a deviated septum;
(e) hypertonus of the orofacial musculature; and
(f) an imbalance between the number or size of teeth and the size of the oral cavity.

The individual stomatognathic system is basically determined by heredity. The relative size and shape of the various portions of the mandible and maxilla, the anatomy of the teeth, the neuromuscular system that governs their functions, and the pattern of growth that the individual elements will follow are all determined by genetic constitution. Forces that undertake to establish and maintain a physiological occlusion must act within the limits established by heredity; the action of the lips and cheeks as they seek to balance the expanding force of the tongue, the muscles of mastication designed to bring the inclined planes of the occlusal surfaces into forceful contact, and the patterns of masticatory movement are thus initially governed by innate tendency. We might well inherit a tendency toward abnormal swallowing. The presence of specific genetic factors in the development of the tongue thrust patterns is, at the present time, however, merely speculative and unsubstantiated by carefully controlled research.

3. **Digit-Sucking.** Research and clinical experience have convinced us that adverse resting postures of lips and tongue are the bases for tongue thrust and are major contributors to anterior malocclusions. Sucking of thumb or fingers promotes an open lip and forward tongue resting

posture. One cannot swallow in a normal manner when the thumb is occupying the space the tongue should occupy. Our strong belief is that digit sucking that is chronic and enduring contributes to the habit of tongue thrust. It probably is not one of the major causes, in terms of number of children who tongue thrust.

Nevertheless, there is some evidence to the contrary. Jann and associates (1964), who studied a group of 358 children from the first through the third grades, found no relationship between abnormal swallowing and digit sucking in this group of children. We do regard digit sucking as a foe to effective treatment of tongue thrust. It is not possible to completely habituate proper tongue postures and movements until the sucking habit has been eliminated.

4. **Open Spaces During Mixed Dentition.** As early as 1927, Rogers reported observing the tongue resting anteriorly in spaces created by missing teeth during the mixed dentition stage of development. Hanson and Cohen (1973) measured the total diastema and edentulous space in each child in their longitudinal research during each of five approximately annual visits during mixed dentition. No significant difference in total space was found between children who retained a tongue thrust pattern or developed it during the years of the research, and those who retained a normal swallowing pattern or developed such a pattern by the age of eight years. In individual cases, however, we would not want to rule out the possibility of the open spaces contributing to the development of the tongue thrust.

5. **Tonsils and Adenoids.** Enlarged tonsils theoretically contribute to the presence of tongue thrust by fostering a low, forward posturing of the tongue, and the adenoids do so by interfering with free nasal breathing. When both tonsils and adenoids are enlarged, mouth breathing is encouraged, which further contributes to a forward resting posture of the tongue. Hanson and Cohen (1973) found a significant relationship between enlarged tonsils and the retention or development of tongue thrust in children. We know of no research, though, that relates inflamed tonsils or adenoids causally to tongue thrust. In our experience, grossly enlarged tonsils sometimes occupy space ordinarily filled with the posterior portion of the tongue, and greatly interfere with the effectiveness of therapy; in such cases we postpone therapy until the condition can be remedied surgically, or until the tonsils atrophy.

6. **Allergies.** Allergies affecting the upper respiratory system are similar in their effects to tonsil and adenoid problems: they tend to promote

mouth breathing and habitual forward resting postures of the tongue. Hanson and Cohen's (1973) research found inconsistent relationships among allergies, palatal form, and tongue thrust. In those children in whom the allergies are so persistent or so severe as to contribute to frequent nasal congestion, attention to them by an allergist or ENT physician is a requisite to the success of therapy.

7. **Mouth Breathing.** Holik (1958) found 85 percent of a group of children who breathed habitually through the mouth to have underdeveloped oral musculature. Watson and associates (1968) describe research that found that children with long, narrow faces, or high, narrow palates have a mean greater nasal resistance to breathing than those with short, wide faces, or low, broad palates. Hanson and Cohen (1973) found a significant relationship between the presence of mouth breathing and the retention or development of tongue thrust in their subjects.

8. **Tongue Size: Macroglossia and Microglossia.** A few patients we have seen appear to have an inordinately large tongue. It is difficult to measure tongue size in a live subject, and even if accurate measurements were obtainable their reliability would be questionable, due to the variability of the size of the tongue in an individual. We believe that true macroglossia is a rare condition, although many deviant swallowers give the impression of having an oversized tongue. When the tongue rests and functions forward and low in the mouth, it is easy to mistakenly call it "macro." Once muscles have been trained through therapy, tongues that seemed definitely large have assumed a much more normal appearance.

Cases have been reported in which the tongue was too small. Ballard and Bond (1960) describe patients in whom the tongue is inadequate to fill the oral space, and in whom a tongue thrust was observed. Patients with tongue-size approaching microglossia have been seen in our offices and are usually difficult, but not impossible, to train.

9. **Orthodontic Treatment.** Patients or their parents occasionally report that there was no tongue thrust prior to the initiation of orthodontic treatment, but that during or following such treatment the tongue very obviously began to push against the teeth. This may be true in selected cases, but in our experience the tongue thrust occurred in most patients since birth and was simply not observed until orthodontic treatment began. This is not to say that such instances never occur. Particularly when there is a severe Class II or Class III malocclusion, while molars are being moved over cusps of molars in the opposing jaw, there is a

period when cusps meet cusps and open spaces are created between upper and lower teeth. Contacts between cusps may make firm occlusion uncomfortable and encourage the tongue to become a pad between opposing sets of teeth. In this manner a tongue thrust pattern might be induced, which could become permanent.

10. **Oral Sensory Deficiencies.** Patients occasionally demonstrate a reduced sensitivity to stimuli in the soft palate and oral pharynx. This has led some observers to postulate a relationship between a diminished gag reflex and tongue thrust. The tongue, they reason, rests and functions more anteriorly in the mouth to take advantage of heightened sensory cues there. Considerable research has been done in the area of oral sensitivity, but most of it has tested the tongue's ability to perceive forms, location, and textures. Small plastic forms are placed on the tongue. Without being permitted to see it or manipulate with the fingers, the subjects must identify the shape or match it with an identical form placed before him in company with other, dissimilar ones. To test two-point discrimination, dividers are set at various degrees of separation to determine the minimum distance between the two points at which the subject is able to perceive separate sensations. Normal subjects can perceive separations of 2 mm or less at the tongue tip.

Ringel (1970) reviews four investigations that find that children and adults with defective articulation have poorer form recognition ability than do normal speakers. There is evidence that measurements of oral form discrimination can differentiate between degrees of articulatory proficiency. Silcox (1969) found no significant differences between performances of tongue thrusters and normal swallowers on oral stereognostic tasks. Bosma (1970) postulates that "the definition of oral functions by their sensory elicitations may apply only to the functions in their nascent state. When stabilized in mature patterns they may acquire autonomy, like that which mature speech seems to possess. At that point, they may be little affected by variations in oral sensation." This may apply as well to nonspeech tongue behaviors.

The serious limitation of oral stereognostic tests, two-point discrimination tests, and texture discrimination tests is that they do not test functions pertinent to sensory abilities of people with tongue thrust. Localization studies do test what should be tested, but the artifact of the pseudopalate may alter perceptions being studied. What do we want to know about the sensory abilities of these patients? We want to know if they can tell whether their tongue is contacting, or pushing against, their teeth, lips, or palate. Do they know when their lower lip is tucked under the upper teeth? If they can perceive linguopalatal contact, can they determine which part of the tongue is touching which part of the palate? Are they capable of sensing direction and degree of movement of the tongue during swallows? Are they aware of whether they have formed a bolus or simply scattered crumbs?

Clinically, we have seen many patients who seem to have sensory deficiencies. Many are unable to tell when the tongue tip is raised, lowered, pointed, or rounded. After repeated practice, some still cannot place the tongue tip on the same spot on the palate twice. Others are unable to sense whether the bolus rests on the anterior or middle portion of the dorsum of the tongue. When we have worked on elevation of the posterior tongue by means of the /k/ sound, a significant number of patients have been unable to tell whether the tongue was making contact with the posterior palate. When we reach the state in therapy where the patient is to swallow food with the lips gently touching each other, some who have swallowed with a labiodental seal previously are not able to differentiate between that kind of contact and bilabial contact. These are some of the sensory skills that practical tests might be designed to study.

11. **A Developmental Theory.** Eating patterns change from infancy through childhood. They begin as a primitive, totally reflexive process and develop into a complex, integrated voluntary-reflexive process. Bosma (1970) referring to the role of consciousness in the developing swallow, states that some functions change postnatally only by the addition of awareness and volition. Into the realm of conscious behavior he tentatively places approximation to a suckle stimulus, the suckling action itself, and the initiation of swallowing by the penetration of the bolus into the pharynx. As the infant develops and progresses to baby foods, junior foods, and, when teeth erupt, to foods that must be chewed, a greater portion of the time occupied by eating is spent in preparatory phases and less in swallowing. Repeated tongue thrusting can occur during the chewing and moving of foods. Until the bolus enters into the pharynx, it is under some degree of voluntary control. Grasping the fork or spoon, directing the food toward the mouth, biting, chewing, forming the bolus — these are learned functions. As you watch people eating in a restaurant you perceive that they do not all eat in the same way. They have *developed* different ways of eating.

The following developmental changes occur during infancy and early childhood:

a. The hyoid descends, lowering and retracting the tongue.
b. The oral cavity enlarges; at rest the tongue no longer contacts the lips or palate.
c. Movements of the tongue, lips and mandible, which occur synergistically, as a composite motor organ in infants, change in response to growth and the eruption of teeth. Biting, chewing, and food-moving, and bolus formation develop new motor patterns for muscles of the tongue.
d. Movements of the posterior pharyngeal wall to assist swallowing diminish. The tongue and velum are displaced to a greater degree, and the pathway of the bolus is more dorsal.
e. Reflexive cueing for the swallowing reflex is supplemented by other cues, such as the contraction of tongue and cheek muscles during swallowing.
f. Mandible and maxillae increase in length, affording more anteroposterior space to the tongue and allowing it to rest more posteriorly, away from the lips, and later, away from the teeth.
g. Growth of the ramus of the mandible, together with the eruption of the teeth, increase the vertical dimension of the oral cavity. These developments provide more space for the tongue, permitting it to effect a seal by squeezing up against the palate, rather than pushing against the lips or anterior teeth.
h. Vertical and anteroposterior enlargement of the oral cavity provides more oral space, hence the oral-pharyngeal airway is opened even though the tongue is held in a more retracted position.
i. The preparatory phases of swallowing, such as biting, chewing, food moving, and bolus formation are cortically controlled. The swallowing process becomes partly voluntary and partly reflexive.

Deterrents to these developments may interfere with the development of normal eating and swallowing patterns in infancy and in early childhood. Some of the factors that may present such deterrents are:

a. The tongue grows abnormally large compared to the size of the oral cavity, creating a crowding condition and fostering the retention of infantile lingual resting postures.
b. Solid and semisolid foods are not introduced into the diet of the child at the appropriate age, and the new motor patterns that

would normally develop to accomplish the biting, chewing, and moving of food do not emerge. Apposition of the tongue, lips, and palate persists during sucking and swallowing. The disruption of developmental timing, together with the additional habit strength due to prolonged infantile feeding patterns, cause the swallowing pattern to remain unchanged when solids are introduced into the diet.

c. The initiatory phases of the swallow, due to the extension of bottle or breast feeding, remain under reflexive control. The development of voluntary muscle contractions of the tongue and cheeks is postponed.

d. Growth of the mandible and/or maxilla is inhibited or retarded, and the tongue is not afforded needed additional anteroposterior or lateral space.

e. Insufficient or delayed eruption of the incisors or molars limits the enlargement of the oral cavity in a vertical direction. The crowded tongue persists in resting and moving forward in the mouth.

f. The nasal airway is restricted in size by the presence of obstructions, such as enlarged adenoids or swollen membranes. Enlarged tonsils partially block the oropharyngeal port. The mandible is habitually depressed and the tongue held low and forward in the mouth to facilitate mouth breathing. The tongue remains forward during swallows.

g. Genetic factors preclude normal size relationships, among the mandible, maxillae, and tongue. The tongue rests and functions principally in the mandibular arch rather than in the smaller maxillary arch.

h. Oral perceptual skills are inadequate, and the normal refinement of movement required for the development of the nonthrusting swallow is not achieved.

None of these speculative conclusions regarding possible early developmental etiologies of tongue thrust has been definitively proven to be valid. Evidence is substantial, however, that alterations in normal physiological development are strongly associated with the retention of the infantile swallowing pattern.

Changes in Later Childhood. Swallowing patterns of most children undergo maturational transitions prior to the period of mixed dentition. They change from an infantile swallow, with the tongue pushing against, or protruding between, upper and lower teeth, to a "mature" swallow, with the tongue tip assuming a position posterior to the upper alveolar ridge during swallowing.

Hanson and Cohen (1973) found considerable variability in maturational changes in their longitudinal study of children ages 4–9 through 8-2. Some who were tongue thrusting at 4–9 learned to swallow normally at some time during the period of the research, whereas others who had demonstrated normal swallow patterns at the initial evaluation developed a thrusting pattern over the years. Of the 103 subjects who at 4–9 protruded the tongue during swallowing, 63 (61.2%) had changed to a normal pattern by the age of 8-2. Of these 63 children, 34 had normalized their patterns between the ages of 6–7 and 8-2. This is the developmental period characterized by the greatest number of missing anterior teeth. When first examined at 4–9, 75 of the children exhibited normal swallowing patterns. Of the 75, 25 (33%) had changed to a tongue thrusting pattern by the age of 8-2.

Conclusion

On the basis of evidence available at the present time, we believe that tongue thrust is a normal behavior in infants. At about the age of five years, more children appear to be swallowing without a tongue thrust than with one. When we combine research results with our clinical experience, we have to believe in a multiple causation theory. We are convinced of the strong interrelationships between form and function. The basic oral anatomy that a child inherits and develops determines to a great extent, for example, the habitual resting position of the tongue. The tongue learns to rest in a place and manner that offer the least physiological resistance. If the tongue rests in a low anterior position in the mouth, approximately once every minute, when a saliva swallow occurs, the tongue will probably move as little as possible to accomplish the swallow. Since any kind of swallow requires a tight seal between the tongue and some part of the mouth, this seal will probably be made anteriorly rather than superiorly. The child who swallows saliva by pushing the tongue against the anterior teeth usually swallows liquids and foods in this manner as well.

Anything that fosters a low, forward resting position of the tongue, then, can contribute to the retention or the development of a tongue thrust swallow. Certainly mouth breathing contributes to such a resting posture. Anything that fosters mouth breathing may be contributing to tongue thrust. In some children allergies affecting the upper respiratory system are severe and persistent enough to make mouth breathing easier than nose breathing. Enlarged adenoids, swollen nasal membranes, and deviated septums are culpable in many children.

The child with a narrow and/or high palatal arch often finds it difficult to rest the tongue in that arch and resorts to a low habitual tongue posture. A lingual crossbite, found in the Hanson and Cohen research to be significantly related to tongue thrust, may be a factor because the resulting relatively greater width of the mandibular arch encourages the tongue to rest low in the mouth. Vertical crowding of the tongue may occur, such as in children with deep overbite. Any of these factors, we believe, may in some children contribute to the abnormal resting position of the tongue and subsequent abnormal swallowing.

REFERENCES

Ardran, G. M, and Kemp, F. H. (1955). A radiographic study of movements of the tongue in swallowing. *Dent. Pract.*, 5, 8.

Ballard, C. F., and Bond, E. K. (1960, October). Clinical observations on the correlation between variations of the jaw form and variations of orofacial behavior, including those for articulation. *Speech Path. Ther.*, 55, 53.

Bijlstra, K. G. (1958). Frequency of dentofacial anomalies in school children and some aetiologic factors. *Trans. Eur. Orthod. Soc.*, 44, 231.

Bosma, J. F. (Ed.) (1970). *Second symposium on oral sensation and perception.* Springfield, Ill.: Charles C Thomas.

Hanson, M. L., and Cohen, M. S. (1973). Effects of form and function on swallowing and the developing dentition. *Am. J. Orthod.*, 64, 63.

Holik, F. (1958). Relation between habitual breathing through the mouth and muscular activity of the tongue in distoocclusion. *Dent. Abst.*, 3, 266.

Jann, C. R., Ward, M., and Jann, H. W. (1964). A longitudinal study of articulation, deglutition, and malocclusion. *J. Speech Hear. Disord.*, 29, 424.

Leech, H. L. (1958). A clinical analysis of orofacial morphology and behavior of 500 patients attending an upper respiratory research clinic. *Dent. Pract.*, 9, 57.

Picard, P. J. (1959). Bottle feeding as preventive orthodontics. *J. Cal. Dent. Assoc.*, 35, 90.

Ringel, R. L. (1970). Oral sensation and perception: A selective review. *ASHA Reports* (No. 5), 188.

Rogers, A. P. (1927). Open-bite cases involving tongue habits. *Int. J. Orthod.*, 13, 837.

Silcox, B. L. (1969). *Oral stereognosis in tongue thrust.* Unpublished doctoral dissertation, University of Utah.

Straub, W. J. (1960). Malfunction of the tongue, Part I. *Am. J. Orthod.*, 47, 596.

Subtelny, J. D. (1965). Examination of current philosophies associated with swallowing behavior. *Am. J. Orthod.*, 51, 161.

Watson, R. M., Warren, V. W. and Fisher, N. D. (1968). Nasal resistance, skeletal classification and mouth breathing in orthodontic patients. *Am. J. Orthod.*, *54*, 367.

Weinberg, B. (1970). Deglutition: a review of selected topics. *ASHA Reports* (No. 5), 116.

CHAPTER 8

DIAGNOSIS AND PROGNOSIS

A "diagnosis" can range from the determination of the presence or absence of abnormal behavior to the detailed description of the history, etiology, consistency, severity, and scope of the problem. If the purpose of the report that results from the diagnostic process is solely to inform a referral source regarding the presence or absence of a treatable problem its content will be limited, in most cases, to a brief description of the problem, a statement concerning its treatability, and an estimate of prognosis. If the report is to be comprehensive, it should provide answers to seven basic questions.

Diagnostic Questions

When a child or an adult is referred for diagnosis and possible treatment for tongue thrust, certain basic questions need answers: (1) Is there really a tongue thrust? (2) If so, is it doing any harm to the speech or to the dentition? (3) Are other myofunctional disorders present? (4) Is the swallow pattern consistent? (5) What treatment is indicated? (6) What structural and functional factors characterize the tongue thrust in this particular patient? (7) What type of tongue thrust is manifested?

1. **Is There Really a Tongue Thrust?** Long (1963) compared the swallows of 25 tongue thrusters and 25 normal swallowers cinefluorographically and found the only significant difference between the groups was in the placement of the anterior portion of the tongue during swallowing. Although other research has yielded significant correlations between type of swallow and various structural and functional criteria, no consistent patterns have emerged. Except in the patients whose tongue consistently and definitively protrudes between the anterior and/or lateral upper and lower teeth, the judgment of the presence or absence of tongue thrust has to involve some subjectivity. Many people with normal dentition rest the anterior portion of the tongue against that part of the anterior teeth closest to the gingiva. Mere contact between the tongue

tip and a portion of one or more of the anterior teeth during swallowing does not necessarily indicate tongue thrust. When there is significant pressure by the tip or blade of the tongue against a major portion of the area of an incisor, cuspid, or premolar, then the diagnosis of tongue thrust is warranted. Most clinicians would agree that the presence of even a severe frontal lisp, if swallowing behavior is normal, does not constitute a tongue thrust problem.

2. **Is the Tongue Thrust Doing Any Harm to the Speech or to the Dentition?** Until more conclusive cause and effect data are available, the answer to this question in every case has to be speculative. There is enough evidence, however, of the coexistence of tongue thrust, malocclusion, and dentalization of linguoalveolar consonants that it is important to attempt to answer this question during the diagnostic process. Research has shown that proper attention to the correction of faulty habitual tongue posturing facilitates the correction of a lisp (Christensen and Hanson, 1981). Most orthodontists agree that the presence of tongue thrust hinders the progress and threatens the permanence of the results of orthodontic treatment. If there is a tongue thrust, however, without an accompanying speech or occlusal problem, we see no reason to prescribe treatment for the behavior. Instead, we ask the patient and the parents or spouse to observe the occlusion carefully during the next six months to two years to determine whether there is any change. If there is, they are to contact the clinician for a reevaluation.

3. **Are Any Other Myofunctional Disorders Present?** These would include digit sucking, lip biting, object biting, cheek biting, lip sucking, lip licking, leaning against the hand or fist, and bruxism.

4. **Is the Tongue Thrust Behavior Consistent?** To assess consistency of swallowing behaviors, it is advisable to observe the patient swallowing saliva, liquids, and solids, each several times. Inconsistencies among the media swallowed or among repeated swallows of the same medium might indicate a transitional pattern.

5. **Is Treatment Indicated?** Most of the patients seen for oral myotherapy are referred by dentists who are aware of the requirements for successful treatment of tongue thrust. Therefore most patients seen for an evaluation are accepted for treatment. For some patients, though, treatment is not advisable. Following are some contraindications to therapy:

a. An extremely mild tongue thrust, with no apparent associated malocclusion or speech defect.

b. An inconsistent pattern of swallowing, with frequent normal swallows.

c. A negative attitude in the child toward therapy and an unwillingness to respond to the clinician's efforts to motivate.

d. A patient who is too young or immature.

e. Need for medical treatment before initiation of therapy.

Particularly important is the elimination of structural problems impeding nasal breathing. Examples of medical steps that can alleviate mouth breathing and provide more optimal conditions for therapy include a desensitization program for allergies affecting the upper respiratory system; reduction of swelling of the nasal membranes; lingual frenectomy; and, in some few cases, removal of tonsils and/or adenoids.

f. Need for orthodontic treatment. If the palatal arch is not wide enough to provide a comfortable resting place for the tongue, and expansion is feasible, it is wise to have this done before therapy is initiated. Also, in many children who have an extreme overjet wherein the incisors are tipped labially, the orthodontist will move the upper incisors lingually with an orthodontic elastic. This often requires only a few weeks, and the posterior movement of the teeth makes it easier for the patient to learn to rest the lips in a proper position and to develop better breathing habits.

Another problem sometimes requiring early orthodontic attention is the correction of faulty molar occlusion. It is difficult to teach a normal swallow when the patient is unable to bite down firmly and naturally at the initiation of the swallow.

g. Any significant change in manner of swallowing or reduction in malocclusion during the past year. If such changes are in the direction of normalcy, it is better to postpone therapy to see whether the habit is correcting itself. Up to approximately the age of eight, many children are still undergoing a change from a tongue thrust pattern to a normal one.

6. **What structural and functional factors characterize tongue thrust in this patient?** Even though certain factors have not been shown by research to be related to the persistence of tongue thrust in a statistically significant manner, one or more of them may be important contributors to the total problem in a *given* patient. Examples are the use of a pacifier, allergies, tongue size, degree of sensitivity of tongue or palate, and psychological problems. If we are to successfully eliminate a habit, we must attend to any and all stimuli and conditions that would offer resistance to the retraining.

7. **What type of tongue thrust is manifested?** By keeping in mind the many variations of tongue thrust, the clinician can avoid overlooking

less obvious conditions. Also, the orofacial behavior you observe should be compatible with your identification of type of tongue thrust.

Preliminary Observations

We should miss no opportunity to study the child when s/he is in repose and unaware of our scrutiny. The observations may take place in the waiting room, and the observer may be the unsuspected receptionist. A plentiful supply of comic books, toys, and other such materials should be considered a part of the diagnostic equipment. Much information can be gathered while the child is distracted by interesting materials.

Facial Posture at Rest

The habitual resting postures of the tongue and lips may be the most valuable bit of information the waiting room supplies; they should be carefully noted.

In the child who breathes nasally and whose lips are competent, the lips are lightly closed or only barely parted; if extreme incisal protrusion interferes, the lower lip is in contact with the lingual surface of the upper incisors to maintain oral closure. If lip closure is achieved only with obvious strain and tension of the facial muscles, it is probably phony and seldom occurs during school or while the child watches television. Tongue thrusters tend to sit with mouth ajar. They often give the impression that there is inadequate labial tissue with which to close the mouth, even though this is usually not the case.

Rest Posture of the Tongue

The great majority of abnormal swallowers, including adults, rest the tongue low and forward in the mouth (Fig. 8-1). Some manage to retract the tongue within the lower arch; some allow the tongue to spread over the lower teeth; some protrude the tongue until it is in almost constant contact with the lower lip. Parents frequently report a series of school pictures over the years, in each of which the prominent feature is the tongue.

Lip Movement During Swallowing

Most people swallow saliva at least every two minutes. Normally the lips are quiet during these swallows. If they tend to tighten noticeably or to protrude from the normal resting posture during swallows, tongue thrust is probably occurring.

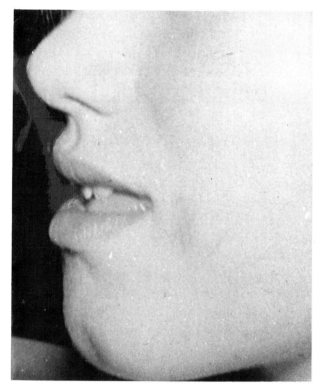

Figure 8-1. Rest posture indicative of abnormal swallowing. Tongue, which appears to be bulging from mouth, proved to be compatible with its oral environment once tongue thrust therapy was completed.

Associated Problems

Watch for signs of other oral habits. The effects of digit sucking may be clearly manifested in the malocclusion. Notice whether the lips are smooth, chapped, or cracked. Some children display an almost continuous series of nonswallow, nonpurposive tongue protrusions.

Detailed Examination

The examination of structures should precede that of functions.

Structures

We will begin with the external structures and proceed posteriorly into the oral cavity.

The Face. The skilled, perceptive diagnostician gleans many clues regarding the client's general health, personality, dental occlusion, and mood of the moment by observing the face. When relationships between upper and lower face, and between soft tissues and skeletal components are in balance, the facial pattern is termed "Class One" (Mason, 1980). Protrusion of the midface area, due either to maxillary protrusion, mandibular retrusion, or a combination of the two, is called a "Class Two" pattern. When the lower face is anterior to its normal position, the face is termed "Class Three."

Ordinarily the plane determined by the lower border of the body of the mandible should roughly parallel the plane created by connecting the upper margin of the ear to the nostril (Mason, 1980). A steep mandibular plane, demonstrated by increased height of the lower face, increases the ratio of anterior vertical oral space to posterior vertical oral space, resulting in a skeletal open bite. The tongue may adapt to this condition by resting forward, against or between the anterior teeth. This may contribute to the development of an anterior open bite.

The Nose. Noses come in a variety of sizes and shapes, most of which provide for free passage of air from the pharynx to the outside world. Most oral myofunctional therapists and speech-language pathologists limit their examination of the nose to a fairly fleeting appraisal of the size of the nostrils, the degree of symmetry, and the patency of the anterior nostrils. Explorations farther back into the nasal chambers are usually done by a physician. An evaluation of function of the nose is more important for the orofacial myologist than a study of its anatomy.

The Lips. The most critical anatomical feature of lips that is of concern to the orofacial myologist is their ability to rest together along their entire length effortlessly. Anterior malocclusions and the necessity for mouthbreathing often create, as years pass, incompetent lips. The upper lip at rest should extend down over approximately one-half, or more, of the surface area of the upper teeth.

Chapping or cracking of the lips may indicate a lip-licking or lip-sucking habit. The lips should be examined for evidences of injury or surgery. If the lips appear to be tight or deficient in tissue, the labial frena should be examined. The lower lip may appear to be extra full, due to its eversion and/or protrusion. An overjet may cause the lower lip to rest against the cutting edges of the upper incisors. A horizontal crease may be present inferior to the lower lip when the overjet causes the lip to evert at rest and during swallows. This crease often disappears spontaneously as orthodontic work corrects the malocclusion.

The Oral Vestibule. Most people have a pair of lateral frena plus a medial frenum inside both upper and lower lips, connecting them to the gingiva. The upper medial frenum is the only one of the six that presents a problem, except in rare instances. The upper labial frenum normally attaches to the alveolar crest at birth but migrates in a superior direction as teeth and alveolar ridge develop; in some cases this migration is incomplete, and instead a tough, fibrous membrane develops between the maxillary central incisors. Surgical excision of this unyielding mass may allow closure of the diastema, although some dentists believe that it should not be done until the permanent canines have completed eruption. Anterior spacing may be a part of normal development, it may be hereditary, or it may be the product of abnormal function of the tongue. If this frenum restricts free movement of the upper lip to any significant degree, a referral should be made to an oral surgeon for evaluation.

The condition of the gingiva should be noted. Inflammation may result in resorption of the gingival margins around the roots of the teeth.

The Teeth. The state of repair of the teeth gives important information regarding the attitudes and interest of the patient and parents. Evidence of poor hygiene should be noted. Intraarch dental deviations include unusually large or numerous diastemata, gaps from missing teeth, excessive crowding of teeth, and asymmetries between the two sides of the dental arch. Abnormal interarch relationships, including open bite, overjet, and overbite, should be measured, not for purposes of defining treatment, but as baselines for movement that often accompanies or follows therapy and precedes orthodontic treatment.

The clinician should be familiar with a classification system, such as the Angle system, since research has found prognostic differences among the various types of malocclusion. Angle Class I patients, for example, have been found to be spontaneously self-correcting with regard to malocclusion as well as to manner of swallowing, more frequently than Class II or Class III patients. Take note of any teeth that do not approximate sufficiently with their fellows in the opposite arch to allow them to function in biting or chewing.

Mason (1980) advises clinicians to count the number of teeth on each side of the midline in each jaw. Any supernumerary or missing teeth will produce asymmetries that will affect interarch relationships and may produce deviations in tongue habits.

It is common for people with overjets, when asked to bite down, to force an alignment of the upper and lower anterior teeth by thrusting the jaw forward. This behavior can usually be averted by asking the patient to bite down on the back teeth.

Since the clinician engages in frequent communication with various dental specialists, a thorough acquaintance with dental terminology is essential. Terms describing various inter- and intra-arch abnormalities should be used appropriately in oral and written communication. These terms were discussed in Chapters Five and Six.

The Hard Palate. The palatal arch should be of a width, height, and conformation that easily accommodates the tongue at rest. Many patients with tongue thrust have palatal vaults that are well within normal limits (Fig. 8-2). Others have a palate that is almost diagnostic in itself. In some instances the palatal arch may almost resemble a cleft. The lateral gum pads may extend so far toward the midline, forming broad shelves on either side, that coaptation of the tongue to the central portion of the palate is a physical impossibility (Fig. 8-3). In other cases, such as the so-called collapsed upper arch, some preliminary orthodontic treatment must be instituted before any hope of swallowing correction can be offered.

More characteristic is the palate seen in Figure 8-4. The anterior region is featured by a marked indentation that is usually circular in form, whereas the posterior palate may be abnormally flat, probably from excessive pressure of the posterior tongue along the distal margin of the hard palate.

To test for adequacy of palatal arch width for accommodating the tongue, instruct the patient to place the tongue tip against the upper alveolar ridge with the mouth open wide, then slowly close the jaws. If the arch is wide enough, the tongue will usually snuggle up within the arch. If it appears to have difficulty fitting into the arch, show the patient how your own tongue does so and ask him/her to imitate you. If the arch is obviously narrow and the tongue appears to have normal muscle tone, yet is unable to fit into the arch, therapy may have to await arch-widening. In many patients it is wise to work on tonicity of tongue muscles before concluding the arch is too narrow.

The Soft Palate (Velum). Examine the soft palate by having the patient sit erectly with the eyes looking straight ahead. Observe velar elevation during the production of "ah" with the mouth open wide. When there is hypernasality in the voice, observe the coloring of the velum; a bluish tint may indicate the presence of a submucous cleft. If the voice is normal, velar length and mobility for vegetative purposes will also be adequate.

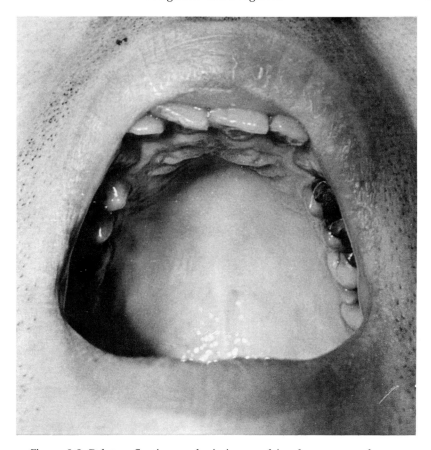

Figure 8-2. Palate reflecting no deviation resulting from tongue thrust.

Mason (1980) provides guidelines for assessing "effective length" of the soft palate. When the velum elevates, note the location of the velar dimple. Normally the dimple occurs at a point about ⅘ of the distance from the anterior border of the velum to its posterior border. If the location of the dimple approaches a point halfway front to back margin of the soft palate, the effective length of the soft palate is so limited that therapy for increasing its motility may prove futile.

The Uvula. Variations in sizes and shapes of uvulas, and even the absence of a uvula, are more interesting than ominous. If the voice is normal you can disregard the uvula. If there is hypernasality a bifid uvula may be another indication of a submucous cleft. The uvula has little significance for disorders of swallowing.

The Tongue. The same generalization made above regarding sizes and shapes of uvulas might be made about the tongue. Some tongues appear

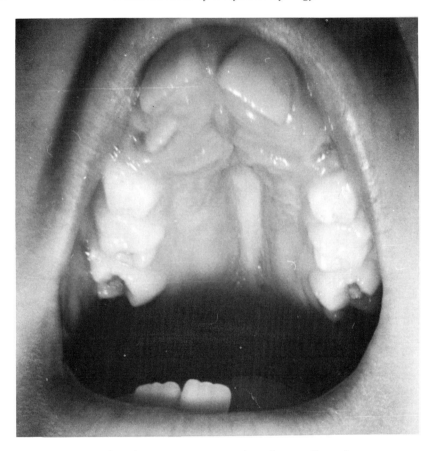

Figure 8-3. Palate that appears never to have known lingual pressure.

to be unmanageably large only to "shrink" as a result of only a few weeks of exercising. Tongues with unusually deep grooves or interesting geographic markings function normally for speech and for chewing and swallowing. If the patient can enclose the tongue effortlessly within the upper arch when the tip is positioned against the alveolar ridge, the tongue can usually be trained to rest in that position and to elevate to that position in swallowing.

Therapists occasionally have patients referred to them for whom the diagnosis of macroglossia has already been made, and surgery to reduce its size already recommended. If the observations made by the clinician warrant the conclusion that the tongue is not excessively large, such a conclusion should be unflinchingly communicated to the specialist who labeled the condition as macroglossia. In some cases a period of muscle training should be recommended and a subsequent referral made back to the specialist for reevaluation.

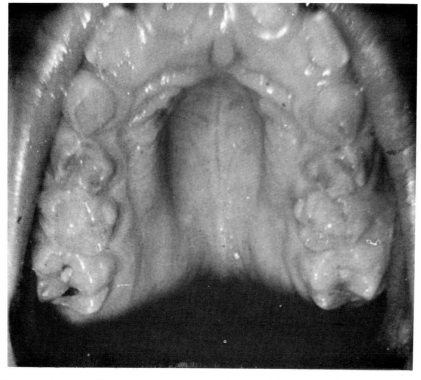

Figure 8-4. Type of palate most commonly found with myofunctional disorders.

Functions

The orofacial myologist deals with four functions of the above-described structures: resting postures, breathing, vegetation, and speech. For therapists who are not speech-language pathologists, knowledge of functions for speech is important for determining necessity for referral to an speech-language pathologist.

1. **Resting Postures.** These have been alluded to in the discussion of preliminary observations. It is not abnormal for patients to rest with lips slightly parted; such a position does not necessarily indicate mouth breathing. Nor do all non-tongue thrusters rest their tongue against the upper alveolar ridge; many rest it within the mandibular arch, and most rest the tongue very close to, or even touching slightly, the borders of upper or lower teeth attached to the gums. If the upper and lower teeth are clearly separated at rest, and/or the tongue is seen to be resting against more than half the lingual surface area of either the lower or upper incisors, such postures are possible threats to the stability of

the teeth. Most patients are able to tell whether they have difficulty keeping the lips together at rest, if the examiner will simply ask the question.

Resting postures of lips and tongue are related to head and neck postures. Saboya (1985) advocates a thorough postural evaluation prior to therapy for oral myofunctional disorders. She particularly studies position of the head on the spinal column, in static position during natural postures, and during all types of movements. If physical therapy is indicated, it is administered prior to the onset of therapy for tongue thrust. If necessary, the speech-language pathologist works in conjunction with the physical therapist.

2. **Breathing.** The person who breathes exclusively through the nose or through the mouth probably does not exist. Most breathe predominantly through one or the other, though, and most patients are able to tell the examiner which is the usual pattern. Many young children honestly do not know, nor do their parents. In that case, ask the child and parents to spot-check occasionally during the week and keep a record for you to examine at the next visit. Observe the client during the consultation. If the lips are parted, ask him/her to try keeping the lips closed for a few minutes, then ask whether it was uncomfortable to do so.

Listen to the patient breathe. If noise seems to come from the throat, it may signify either congestion in the pharyngeal airway or tension accompanying breathing. If it comes from the nasal passages, it is indicative of restriction, from unusually small nares, swollen membranes, a deviated septum, enlarged adenoids, or the presence of mucus. There is little the orofacial myologist can do, except to encourage the patient to try to keep the lips closed and breathe through the nose, or assign the more frequent use of a handkerchief or tissue. A referral to an otolaryngologist is always a safe procedure to follow in case of doubt.

3. **Vegetation.** This term, in the absence of a more appropriate one, deals with all the activity preceding, accompanying, and following swallows of food, liquids, and saliva. In each case we will describe abnormal behaviors that may or may not be part of the pattern of the individual patient.

Food. The tongue reaches out for the food as it approaches the mouth. A large bite is taken. During chewing the lips are either constantly or intermittently parted, and the tongue presses against the anterior teeth or protrudes between upper and lower teeth. Chewing may be very brief or continued for a prolonged period. Swallowing may occur several times during one mouthful of food. Immediately prior to, and during, swallowing the lips may protrude and/or compress noticeably. After the swallow the patient may be seen to clean up food that has remained and swallow again. As the lips part following the swallow, the tongue remains against, or is seen to leave, the anterior teeth. An inspection of the tongue reveals an abnormal amount of food remaining on its dorsum.

During swallowing, palpation of the masseter muscle may reveal little or no contraction. After checking food swallows without interfering with lip activity once or twice, break the labial seal as the swallow begins, to see whether the tongue is against the front teeth. One way to do this is to position a finger lightly on the thyroid notch of the larynx and the thumbs on the upper and lower lips. Tell the patient to swallow. As the larynx begins to raise, quickly separate the lips. In the case of a deep overbite that obscures the tongue, place a tongue depressor the flat way between the upper and lower lateral teeth just prior to the swallow. This does not usually produce a tongue thrust in a patient who does not already have one.

Liquids. The head is tipped forward and the tongue reaches forward as the glass nears the mouth. The lips are active during continuous drinking. As the lips leave the glass following the last swallow, the tongue may be seen to leave the front teeth. Have the patient take a sip and hold it until you say to swallow. Part the lips in the manner as for food swallows, making sure you don't part them too soon. The tongue thrust, if present, is easily visible.

Saliva. Ask the patient to nod the head when there is sufficient saliva ready to swallow. Part the lips as for food and liquid sipping. If repeated attempts are necessary, squirt a little water under the tongue to simulate saliva handling.

In some cases great resistance is met when the thumbs attempt to break the labial seal during swallows. When this occurs the patient almost invariably has a strong tongue thrust. Repeat the attempt two or three times to determine consistency of pattern. Timing is a critical factor in breaking the lip seal, and the novice will need experience with dozens of patients before feeling competent.

4. **Speech.** A complete speech evaluation is probably unnecessary. We find that engaging the patient in conversation provides ample opportunity to check for the presence of a lisp or the dentalization of linguoalveolar sounds /l/, /t/, /d/, /n/. Have the child count to twenty and watch tongue

contacts from a side angle. If any doubt concerning the normalcy of speech remains, a formal articulation test may be administered.

Statement of Diagnosis. In addition to declaring whether or not a tongue thrust is present, the examiner should define the type of tongue thrust and, in some manner, assess the severity and consistency of the abnormal swallow. We use a 0-1-2 rating scale. If the tongue does not contact any part of the anterior or lateral teeth during swallow, a rating of "0" is assigned. If there is significant linguodental contact, but only against the lingual aspect of the anterior or lateral dentition, "1" is assigned. If the tongue protrudes beyond the cutting edge of any of the upper or lower anterior or lateral teeth, excluding the molars, the rating is "2." The "0" translates, in written reports, into "no tongue thrust," the "1" into "mild" or "moderate" and the "2" into "severe."

Any special problems or positive considerations should be noted, including an estimate of the motivation of the patient and parents. A prognostic statement is always appreciated by the referring dentist.

Recommendation. If therapy is recommended, is it going to begin soon? If the clinician sees the need for possible surgery, orthodontic treatment, or attention by any other specialist, s/he should so indicate. The decision concerning therapy or alternative steps to be taken by the patient or parents should be clearly explained and substantiated by the clinician. There should be no doubt as to what steps should be taken next and by whom. If therapy seems to provide a logical precaution in a given case, it should be entered into with gusto.

Prognosis. Two questions may be answered: (1) Is the tongue thrust likely to self-correct with time? (2) What is the probability of success of therapy? Answers that depend on anatomical and physiological considerations are similar. Children under the age of seven years who have very slight malocclusions, with Class I molar relationships, who breathe habitually through the nose, whose palatal arches are ample in width and normal in configuration, and whose thrusting behaviors are inconsistent are more likely to spontaneously change swallow behaviors, and also more likely to respond favorably to therapy. Add to these factors maturity, motivation of child and parents, normal intelligence, a positive attitude of the referring specialist as conveyed to the patient, and an activity schedule not overly busy, and the prognosis for therapy is even more favorable.

Summary

A complete diagnostic evaluation should include a case history and a thorough examination of oral structures and their functions in swallowing and nonswallowing activities. Observations of the patient's oral behaviors are more valid when the patient is unaware of being observed. Random, unsolicited saliva swallows are more meaningful diagnostically than swallows done on command.

The assessment must include a thorough examination of all aspects of the tongue thrust. The consistency of the patterns should be determined by repeated trials on swallows of food, liquid, and saliva. The orofacial myologist need not make the diagnosis without help. The patient, parents or spouse, the dentist, various medical specialists, and teachers are sources who can provide diagnostic information.

Certain characteristics have been found to be important prognostically. Enlarged tonsils, a high and/or narrow palatal arch, mouth breathing, and lingual crossbite appear to be significantly associated with the retention of a thrusting pattern through the mixed dentition period of development. The total diagnostic picture serves as a basis for determining habituation, or regression, during or after treatment. Individualized therapy planning is made possible when careful attention is paid to all aspects of the oral behavior. A thorough examination will permit the exclusion of unnecessary exercises and assignments and the inclusion of pertinent ones. This first step in treatment should not be done hurriedly.

REFERENCES

Christensen, M, and Hanson, M. (1981, May). An investigation of the efficacy of oral myofunctional therapy as precursor to articulation therapy for pre-first grade children. *Journal of Speech and Hearing Disorders, 46*(2) 44–46.

Long, J. M. (1963). A cinoflurographic study of anterior tongue thrust (Abstract). *American Journal of Orthodontics, 49,* 865.

Mason, R. (1980, April). Principles and procedures of orofacial examination. *The International Journal of Oral Myology, 6*(2), 3–16.

Saboya, B. de A. R. (1985, July). The importance of the axis in the study of oromyofunctional disorders: an integrated approach. *International Journal of Orofacial Myology, 11,* 5–13.

CHAPTER NINE

A REVIEW OF TREATMENT APPROACHES

There are many ways to cope with tongue thrust. It can be ignored. The orthodontist can straighten the patient's teeth and wait for the tongue thrust to disappear when the malocclusion is corrected, or the orthodontist can fasten an appliance to the lingual aspect of the teeth to encourage the tongue to stay away from them. It can be treated surgically. Hypnosis can be employed in its treatment. Oral myotherapy is another alternative; it can be perfunctory or comprehensive, and it can be administered intensively or extended over a period of years. Some approaches combine the use of appliances with myotherapy. Still another approach combines surgery with a reminder appliance.

Representative treatment programs will be presented in this chapter. Some are programs that have been tried in the past and abandoned. Others are currently in use.

Appliance Therapy

The use of "habit appliances" is the historic approach of dentistry to tongue thrust. They are still used by some orthodontists, and oral myofunctional therapists should be able to recognize them. They come in infinitely assorted styles, a few of which will be described, together with the philosophy behind their use.

The Hay Rake. The hay rake appliance consists of a metal bar or wire fitted to the lingual surface of the maxillary incisors. It is usually held in place by welding the ends to metal crowns over the first permanent molars or the second deciduous molars, and is equipped with a row of prongs, usually four to six in number, welded at right angles to the bar and projecting posteriorly and downward. One example is seen in Figure 9-1. These prongs are frequently sharpened and are positioned in such a way that the tongue may be protruded only at the expense of laceration. Some have been seen in which the prongs were placed in light contact with gum tissue above the maxillary incisors, and a loop of wire

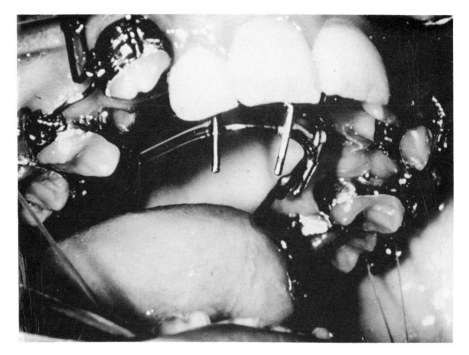

Figure 9-1. One of many types of hay rake construction.

was so placed behind the teeth that protrusion of the tongue caused
torquing, driving the needle points into the gums.

The entire philosophy behind the hay rake is to make tongue thrust so
painful that the patient will stop the behavior. It does not teach him/her
to swallow correctly, and the resulting patterns of deglutition may be
quite bizarre. Not surprisingly, many patients learn to avoid lingual
protrusion; as might also be expected, their resultant substitute may be
as harmful as the original thrust. It is now becoming common to implant
the prongs in an acrylic plate that is adapted to the lingual surface. In
many cases the prongs extend upward and backward from the mandibular arch, making impossible any normal function of the tongue in the
anterior portion of the mouth.

The Cage. The cage is an extension of the hay rake; its prongs are not
sharpened but are much more numerous, usually ten to twelve, and are
sometimes an inch long. The cage is welded to either the upper or lower
arch, the distal ends of the wires hanging free. Although more humani-
tarian in concept, the cage frequently results in even greater problems.
The child is asked to place the tongue within this wire cage, thus preventing
lingual excursion between the teeth. Provided the child cooperates, the

added resistance provided by the cage serves only to confirm and strengthen the pattern of protrusion in lingual muscles. Regardless of whether the upper or lower arch is employed for anchorage, the cage renders marked insult to speech, since normal articulation of several consonants is obviously impossible. One 14-year-old girl was seen after she had been through orthodontic treatment twice. She had worn a cage over a year, but her bite was more open than in the original malocclusion and her speech was grossly distorted. This child was dropped from therapy, to the great sorrow of all, because she had become convinced that everything was hopeless and that she was a failure; she refused to make any further effort.

The Fence. Used only on the maxillary arch, the fence, also called the crib, begins as something similar to the hay rake. A heavy arch wire is fitted, in this case immediately posterior to the gums, and welded to molar crowns. A wire screen is suspended vertically from the arch wire; this screen may take the form of another piece of arch wire bent into a series of broad V shapes, the wires may crisscross, or the crib may be a thin sheet of acrylic. The tongue encounters the fence before it arrives between the teeth and is thus harmlessly contained without pain. The dentist who installs a fence should be aware that the most effective method by which to strengthen a muscle is to provide resistance for that muscle. As with the cage, the end product of a crib, once it is removed, is a tongue with an enhanced capability for tongue thrust.

Cleall (1965) placed cribs in the mouths of 20 patients and then compared cinefluorographic films of each adolescent before the crib was placed with those made immediately after removal of the crib six months later and others made after an additional two months. He found that the tongue resumed much of its original thrusting as soon as it was free of the crib, and that after two months of regained freedom it had reverted almost entirely.

Probably the best that can be said for a fence is that it interferes considerably less with speech. In combating tongue thrust, it restrains; it does not retrain.

The Curtain. The curtain consists of a U-shaped sheet of acrylic that drops directly down from the palate, usually at about the cuspid region; by being somewhat posterior to the incisal ridge, it may extend a bit lower in the mouth. It is often braced by an acrylic palatal arch that in turn is held firmly in place by continuous loops of arch wire extending around the buccal surface of the molars. The reasons given for its use are the same as for the fence, but its function is somewhat different. It might actually restrain the tongue if tongue thrusters really close their teeth before swallowing; the more common sequence is: The tongue is protruded, such closure as may occur is initiated—and the curtain comes down across the dorsum of the tongue like a Spanish bit in a horse's mouth, effectively keeping the teeth apart even farther than usual.

Reminder Appliances. Some orthodontists advocate the use of a "reminder appliance," rather than the more tortuous rake or crib. For example, Walker and Collins (1971) suggest anchoring a transpalatal reminder wire to the maxillary cuspids. The patient is instructed to keep the tongue behind the wire when he talks, eats, and swallows.

A different type of reminder is described by Littman (1968). A pear-shaped opening, extending from the distal aspect of the canines to the mesial aspect of the second premolars, is made in an upper Hawley retainer. The patient wears this appliance at all times. He places the tongue tip in the opening, bites down on his posterior teeth, and swallows. The tongue tip is to remain in the opening after each swallow. He practices this exercise twenty-five times, three times each day. The usual treatment time is approximately three months.

Another modification of the Hawley appliance is suggested by Kaye (1973). Palatal extension loops are attached to a retainer. The loops extend inferiorly to the lingual cervical areas of the lower incisors. They reportedly restrict tongue movement only anteriorly, allowing free movement laterally and vertically. This is similar to the "fence" described earlier in this chapter, except that the Kaye appliance is removable. Kaye instructs his patients to remove it only for meals and tooth brushing. An average of six months wearing time is required for most patients to eliminate tongue thrust, according to Kaye.

The Andresen Appliance. Some orthodontists advocate treatment of tongue thrust with the Andresen appliance. This device completely encircles the lingual aspect of both dental arches, forcing the tongue away from the teeth. The patient swallows with his tongue restricted to the small space available to it. In patients with "functional" open bites, the appliance frees the teeth from the tongue's intrusive influence and allows them to erupt to a normal bite. Most writers recommend supplementing the use of the appliance with myofunctional therapy.

The Andresen appliance also can be used in the treatment of distocclusion related to tongue thrust. As the patient swallows with the appliance in position, the swallowing muscles cause the upper dental arch to move distally, while the lower arch is encouraged to move mesially, it is claimed.

The Bionator. Balters' Bionator is similar to an activator appliance, but is less bulky and allows the child to speak normally with the appliance in place. It does not cover the anterior palate, and fits loosely in the mouth. It can be worn at all times, night included, except during eating.

There are three types of Bionators: the standard (for Class II, Division 1 malocclusions), the Class III, and the open bite appliance. Essential elements for all three types are a vestibular wire and a palatal arch. Lip closure is necessary for the treatment to be effective.

The purposes of treatment, according to Graber and Neumann (1977) are: (1) to promote lip closure and contact between the back of the tongue and the soft palate; (2) to enlarge the oral cavity; (3) to move the incisors into an edge-to-edge relationship; (4) to elongate the mandible, allowing for more favorable tongue resting postures; and (5) to improve relationships among the jaws, tongue, dentition, and surrounding soft tissues. Although Balter's explanations seem logical, the permanence of the restraint forced upon the tongue by the appliance is questionable.

The Oral Screen. Graber and Neumann (1977) contend that the oral screen, primarily designed as a "lip molder," is also helpful in the correction of tongue thrust. A double screen, one on the labial aspect of the anterior teeth, the other on the lingual aspect, keeps unwanted tongue and lip pressures away from the teeth. Its most helpful contribution, in our opinion, would probably be to promote nose breathing, allowing the tongue to be trained to rest "on the spot." It is used principally on the early deciduous dentition.

Advantages of Habit Appliances. One of the perplexing problems in the field of oral myofunctional disorders is that so many treatment procedures are successful with *some* patients. Some orthodontists today still routinely use rakes to overcome tongue thrusting. Had they not been successful with this procedure in a significant number of patients, they undoubtedly would have discarded it long ago. From discussions we have had over the years, though, we would have to conclude that those orthodontists who have had extensive experience with appliances and with myotherapy prefer the latter. Certainly the use of the appliance keeps the patient's progress more directly under the orthodontist's control. He does not have to worry about the qualifications, personality, or general effectiveness of an oral myotherapist. The appliance is concrete and stable. What, then, are its disadvantages?

Disadvantages of the Anti-Tongue-Thrust Appliance

1. It impresses many parents and children as being a type of torture device.

2. Habits developed with it in place are dependent on the cues, or feedback, it provides when the tongue is misplaced at rest or moves forward during swallowing. When the appliance is removed, these cues are withdrawn. The tongue's environment again more closely resembles its former state, when the thrusting habit was present. There is a strong tendency for old habits to return.

3. Its presence often makes proper production of sibilant sounds very difficult. The resultant lisp is sometimes only temporary but often persists until the rake is removed. This is a source of embarrassment to the patient and can affect his social verbal behavior and his self-concept.

4. The teeth can be damaged by its use. With regard to cribs and curtains specifically, it has been noted both by Cleall (1965) and by Subtelny and Sakuda (1971), that the effect on the lower arch, in particular, can be unfortunate. Since these appliances withhold all lingual pressure from the anterior teeth, the frequently hyperactive lower lip is enabled to exert unopposed force on the lower arch. The mandibular teeth move lingually, and crowding often results.

A malocclusion is at all times in a state of functional equilibrium; the dangers of disturbing that balance without making compensation are clearly illustrated in this situation. The loss of precious intercuspid distance alone would result in a poor bargain even if the tongue thrust were eliminated, which is usually not the case.

We prefer the use of such an appliance only when oral myotherapy is contraindicated, such as when the patient is mentally or emotionally not equipped to carry out the assignments in a retraining program. If no myotherapy is available, the rake or crib may be better than no treatment at all. Very infrequently, it can be helpful in keeping the tongue away from the teeth until the palatal arch can be expanded enough to make myotherapy more feasible.

If such a device is used, it is helpful if the orthodontist removes it in stages, removing first the spikes but leaving some kind of crossbar in place for a time. If the crossbar can be reduced in size gradually, this is desirable.

Surgery

We have heard reports of greatly improved tongue function, for speech as well as for swallowing, after the surgical removal of a triangular wedge from the anterior part of the tongue. In addition to this wedge, Lines and Steinhauser (1974) suggest the removal of an oval segment from the central portion of the tongue, leaving the lateral margins intact but reducing the tongue in both length and width. Still another surgical procedure involves the detachment of the genioglossus muscle from the mandibular mental spines. This operation, described by McWilliams and Kent (1973) was performed on 7 patients, all with an anterior open bite and tongue thrust. The genioglossus muscle makes up the body of the tongue and is chiefly responsible for its protrusion. Measurements of protrusive tongue forces were taken twice before surgery and four times postsurgically at ten days, six weeks, three months, and six months.

They report that their patients had difficulty swallowing immediately after surgery but felt comfortable doing so 48 hours afterward. The mean lingual pressure recorded presurgically was 4.83, which was reduced to 2.67 after surgery. At the six-month postsurgical recording the mean measurement was 3.22. No orthodontic treatment was instigated, and six months after surgery there were no significant changes in cephalometric and/or plastic model measurements. The authors concluded that detaching the genioglossus muscle significantly reduced the maximum protrusive tongue forces.

The use of this surgical approach as a principal means for tongue thrust and open bite correction must be questioned. The possible usefulness of genioglossus muscle detachment in conjunction with orthodontic treatment or other surgical procedures should be explored. Surgical detachment may be indicated to effect a stable result and to prevent regression.

Hypnosis

Barrett (1967) has described the successful use of hypnotherapy with 25 "failure" patients in his own practice. Several other writers have described hypnotherapy they use with their patients.

In 1965 Crowder listed three types of treatment—mechanical, myofunctional, and psychological—and stated that hypnosis represents the third type of treatment. His hypnotherapy typically involved a series of four appointments spaced approximately one or two weeks apart. At the first appointment impressions were taken of the malocclusion. An explanation of the therapy procedure was given to the patient and to the parents. The patient was then taught to swallow correctly, without the use of a trance. At the second session the child was given an explanation regarding hypnosis and its role in retraining the swallowing. The child imagined a favorite program or movie he had seen. This induction process was the "sleep-game" technique. During the third and fourth sessions a deeper trance was achieved, and strong posthypnotic suggestions were given to the patient. S/he was then seen at approximately two-month intervals for the next six months. At these follow-up sessions, reinforcement of the posthypnotic suggestions was given if necessary, but according to Crowder it was not often required.

A similar technique was reported by Secter (1961), who had a child visualize an unpleasant theater scene during a trance, and then through suggestion produced an association between both tongue thrusting and nail biting and the unpleasant feeling the scene aroused.

Lambert (1959) reported a study involving 12 subjects. The subjects received an average of ten sessions involving hypnotherapy, and of the 12, 10 stopped tongue thrusting.

Palmer (1961) reported on 13 patients with whom he had spent an average of 1 hour 21 minutes in a total of five sessions. His finding was that the depth of the trance was not a critical factor in success in therapy. There were no relapses found in 315 days.

Barrett's (1967) personal experience with hypnotherapy has been very positive. He does not use it routinely, however, but only with patients who have special problems and only as an adjunct to the later stages of therapy. Hanson has tried hypnosis several times with tongue-thrust patients but prefers not to use it routinely. Both prefer the rapport gained in a simple person-to-person relationship in therapy. We have the following objections to hypnosis in therapy for oral myofunctional disorders:

1. Some subjects require several trance inductions before success is achieved. Although children of the ages of most of our patients are generally good subjects for hypnosis, it has not been easy for us to spot the difficult ones.

2. Many hypnotherapists provide little or no instruction in learning of normal function at the conscious level, prior to trance induction. The complexity of the change from a thrusting pattern to a nonthrusting one makes any expectancy of success with this procedure seem to us to be highly unrealistic.

3. Many parents have antiquated apprehensions about hypnosis. We prefer to use the time it would take to correct these impressions in motivating or training the patient.

4. There is no controlled research attesting to the efficacy of hypnotherapy for this problem.

Oral Myofunctional Therapy

In the United States the treatment approach that has gained widest acceptance is myotherapy. Walter Straub's (1962) pioneering efforts in this field were alluded to in Chapter Nine. His was the earliest therapy (1) to provide a complete, step-by-step program for the correction of tongue thrust, and (2) to be published in detail and taught as a method to other professionals. It has served as a model for many programs developed subsequently.

1960–1980

Straub Program (1962). We will begin with Straub's program, and proceed in a loosely chronological fashion through the 1960's and 1970's, then describe some newer programs.

Faced with a lack of information from research on the relative importance of the various muscles and structures involved in the swallowing process, Straub devised a comprehensive therapy program in an effort to omit no possible contributing factor. His program included speech exercises, muscle-strengthening exercises, and habituation assignments. He modified his program often. We present here a summary of one he used relatively late in his development of procedures.

Before beginning the swallow retraining, the therapist helps the patient to eliminate any harmful habits he may have, including "the leaning habit," stomach sleeping, nail biting, pencil biting, and thumb sucking. Work on the tongue-thrusting problem begins with a thorough oral and speech evaluation and an explanation of the problem and its therapy to the patient.

Lesson 1
1. The differences between normal and tongue-thrust swallowing
2. Proper elastic placement on top of tongue for placement practice
3. Proper occlusion of teeth for swallowing
4. Learning to suck when swallowing.

Lesson 2
1. Tongue-popping, or clicking, exercise for placement of middle part of the tongue
2. Swallowing with this part of the tongue in position
3. "Choo" exercise, with teeth together.

Lesson 3
1. Explanation of rest position of tongue and teeth
2. Word exercises, involving /t/, /s/, /d/, and /n/ sounds

Lesson 4
1. "2-s" exercises—consisting of two parts: the "spot" and the "squeeze"
   a. The "spot"—tongue in position for initiating /ch/ sound, which is proper rest position for tongue
   b. The "squeeze"—tongue squeezed tightly against "spot" several times

Lesson 5
1. "4-s" exercise—placement of tongue on spot; salivation; squeeze; and swallow

Lesson 6
1. Practice of /unka/ sound to create awareness of posterior tongue
2. Practice of /k/ and /g/ sounds

Lesson 7
1. Pushing (by child) of tongue depressor upward with tongue
2. More /k/ practice

Lesson 8
1. Two types of whistling—with lips and with tongue
2. Lateral lip-stretching exercise

Lesson 9 (important for child with open bite or short upper lip)
1. Stretching of upper lip over upper teeth
2. Holding of card between closed lips

Lesson 10 (to eliminate crevice caused by overdeveloped mentalis)
1. Puffing out of lower lip by blowing against it

Lesson 11 (important for child with open bite or lateral thrust)
1. "Slurp" exercise—spread the lips wide apart; close teeth tightly together; suck air back vigorously as tongue is drawn back; then swallow

Lesson 12 (for "side thruster")
1. Swallow is accomplished with sharp instrument extending into oral area at premolar area, with care being taken to avoid touching instrument with tongue.

Lesson 13
1. Swallowing of a little water by biting down and sucking the tongue up
2. "Straw-pull" exercise—child holds a straw against roof of mouth with tongue; lips and teeth are apart; he sucks and swallows
3. Practicing saying "kick" and similar words

Lesson 14
1. Yawning practice to stretch throat muscles and pull tongue into back of throat
2. Reviewing of sounds produced by tip, middle, and back of tongue

Lesson 15
1. Gargling taught, so that throat muscles will be strengthened and the tongue pulled downward and backward
2. Elimination of facial grimaces during swallow
3. Patient shown how the hyoid bone moves during swallowing

Lesson 16
1. Exercise of the uvula (raising and lowering)
2. Swallowing with fingers touching lips to check for movement during swallowing

Most oral myologists whose programs are adaptations of Straub's have shortened the part of the program devoted to formal lessons and exercises. Many continue to see the patient for rechecks during the entire period of orthodontic treatment, but few programs we have seen include as many structured lessons as his.

Abbreviated Program (1964). Some clinicians have gone to the other extreme, providing only a brief instruction period and expecting habituation to occur spontaneously. This procedure has sometimes been used by dentists who, not having access to an orafacial myologist, have made at least a token effort to make their patient conscious of the importance of keeping the tongue away from the teeth. One such program is described by Whitman (1964), who first locates the "spot" with an instrument. He then places a broken-up mint on the dorsum of the tongue. The child holds the mint against the spot as he swallows. This exercise is done "constantly" for three or four days. The patient is then advised to take smaller mouthfuls of food, and to swallow the food in the same manner as he swallowed the mint. Whitman attributes this basic method to an old French dental textbook published in about 1840. In those days, he said, they used buttons, Sen-Sen, or wax instead of mints.

We have seen few children who benefited from this limited type of training. We have also attempted this kind of abbreviated therapy with patients who seemed unusually intelligent and well motivated, or who lived so far away from the clinician that further instruction was not possible. In nearly all of the cases it has not proved to be effective. Swallowing correctly is deceptively easy to teach but difficult to habituate, and many clinicians have fallen into the trap of providing too short a training period for their patients.

Mini-Abbreviated Program. Tepper (1986) describes the correction of tongue thrust in one easy lesson. The therapist, presumably the dentist, first teaches correct positioning of the tongue and teeth. "Correct positioning of the tongue cannot be done intraorally so the tongue is actually placed in correct position extraorally in this manner: The patient or operator places the thumb on the hyoid bone with a firm but gentle pressure, pushing the tongue upward and backward into its proper place." The patient is then asked to swallow, as pressure against the hyoid is maintained. Placement of the thumb on the hyoid miraculously accomplishes all sorts of things: the suprahyoid muscles are all strengthened, the teeth are automatically occluded, the tongue is forced upward and backward, the front part of the tongue is automatically placed in the correct position, and tongue muscles function in their proper manner during swallowing. After several tries with the thumb pressure, the patient swallows without the thumb against the hyoid. Carry-over assignments are given, and the problem is taken care of. No further comment seems necessary.

Phonetic Placement Method (1968). Many orthodontists, including Straub, who found it impossible to provide therapy for all of their patients, turned to speech pathologists because of their familiarity with oral anatomy and physiology and their training in the modification of habits. It was natural that the speech pathologists would utilize speech principles and tongue placements and movements in their training programs. One such person was Goda (1968). His approach was to divide the therapy into five goals, as follows:

Goal 1
The child understands the swallowing process.

Goal 2
The tongue tip is elevated to the spot. The child repeats the /l/ sound several times, then the /t/ sound, to get the feel of the seal, and the /d/ sound. The child drills on a series of one-syllable words.

Goal 3

Tongue-tip placement for swallowing saliva and liquids.

An appealing feature of Goda's therapy is the brief, succinct descriptions he teaches the child to remember as he swallows. For liquids and saliva, the three steps are "tongue up, tight seal, swallow." While this goal is being achieved, the child continues to perform the /t/, /d/, and /l/ word drills.

Goal 4

The child occludes the lips and the teeth. Two new steps are introduced, and the process becomes "food in, chew food, food back, tongue up (seal), and swallow." The child drills on sentences and phrases, including the /t/, /d/, and /l/ sounds.

Goal 5

Carry-over of tongue-tip placement for all swallows and for speech. The child is instructed to do much self-observation. Reminders are provided, and a checklist is used to ensure compliance with the assignment.

Goda states that this goal is accomplished after ten to twelve weekly meetings, which we understand include all meetings involved in reaching the five goals.

There are some advantages to this type of approach. Those children who do habitually dentalize their linguoalveolar consonants often rest the tongue habitually against the teeth. Attention in therapy to proper habitual rest position and to correct production of these consonants is often complementary. The speech sounds are something the child readily understands, and they provide a good reference and reminder for him. We include attention to speech training for those clients in whom we find the speech to be defective.

There are also inherent disadvantages. Many children with tongue thrust have no accompanying speech problem and must spend considerable time on word drills needlessly. Attention to proper production of the words is often given at the expense of attention to movements and postures involved in proper swallowing. We prefer to work directly with the swallowing movements and resting postures of the oral structures, providing articulation therapy only when articulatory defects are present.

Program for Young Children (1973). Pierce and Warvi (1973) have written a manual designed to help the myotherapist work with children between the ages of 4 and 9 years. We quote from the instructions to the therapist:

> It is our intention that you use this manual as your guide in correcting the deviate swallowing pattern and that you be flexible about supplementing the program with whatever additional exercises the individual child needs for associated oral habits.
>
> Lessons 1 to 6: Muscle Training
> Lessons 7 to 10: Swallow Training
> Lessons 11 to 14: Habituating the Correct Swallow
>
> It has been our experience that this program can be used successfully on the following schedule: Once a week for 10 weeks (Lessons 1 to 10); once a month for four months (Lessons 11 to 14). After completing the 14 lessons, the child should be seen at six-month intervals for two years.
>
> We have also found that this program can be used successfully in either individual or small group therapy.

This manual uses cartoon-type illustrations to teach and motivate the child. The lessons progress in easy steps, and the successful completion of the lessons requires several months' work. The manual is filled with motivational helps and interesting reminders for the child. After the final lessons are completed, the child is put on a maintenance type of program and is seen periodically for rechecks. S/he is given a special diploma when s/he completes the course.

This manual has been written by people who understand the needs and limitations of children. The language used is simple and entertaining.

Research has indicated that some children between the ages of 4 and 7 years will probably continue the tongue-thrust habit into adulthood. Others will spontaneously develop a nonthrusting pattern. We favor therapy for those children in this age range who (1) have severe malocclusions apparently related to tongue thrust (open bite, overjet, overbite, lingual crossbite), (2) are chronic mouth breathers and demonstrate a severe thrusting pattern, (3) reveal a progressively severe malocclusion with the passage of time, and (4) present a number of related abnormal oral habits (thumb sucking, lip biting, tongue thrust, etc.).

Behavior Modification Therapy (1960's to 1970's). Behavioral psychologists have developed an approach to therapy for behavioral change that gives the child immediate, attainable rewards; many clinicians have found it to be very helpful in their work with tongue thrust and related problems. Behaviorists have defined learning principles and devised an approach based on these principles that alters behavior in a predetermined, highly structured manner. We will not discuss behavioral *theory* in this chapter but will make an effort to describe enough basic *principles* and *procedures* to enable us to show their application in therapy for oral myofunctional problems.

Operant Conditioning. Conditioning is the process in which a behavioral pattern is repeated enough times under certain conditions to enable an observer to predict its occurrence whenever those conditions occur.

Consider, for example, the infant cry, which, of course, is a reflexive response, not a conditioned one. As infants develop, however, they find that certain things may predictably follow their crying response. Their mother may enter the room, pick them up, give them a hug, and hold them for a long time. Since the mother's appearance occurs immediately after (contingent on) the infant's cry, babies learn that they can make their mother appear by crying. They learn, in other words, that a certain response pattern causes changes to occur in their environment. When they use that behavior to bring about those changes, they are "operating" on their environment, and the behavior is called "operant behavior."

In operant conditioning the response occurs first and is followed by a stimulus that encourages or discourages the repetition of the response.

Reinforcement and Punishment. When the stimulus, or change in the environment that follows the individual's response, is of such a nature that it encourages the response to be repeated in the future, the response is said to be reinforced. When the stimulus discourages the repetition of the response, it is called punishment. A given stimulus may be a punishment to one child and a reinforcer to another child, or a punishment or reinforcer to the same child under different conditions. Consider again the example of the baby crying. If the child has been grossly neglected and finds that the cry will bring the attention of the mother, the child may repeat the cry even though the mother shouts menacingly as she approaches the child. The shouting and the appearance of the mother occur simultaneously, and even though the shouting is unpleasant, it is better than not having the mother there at all. The shouting, then, becomes associated with a desired stimulus and encourages the crying response to recur. It is therefore a reinforcer, not a punishment for the crying behavior.

Positive and Negative Reinforcement and Punishment. Positive reinforcement is the presentation of a desired stimulus contingent on a response. Examples are a smile, a nod of approval, a piece of candy, a "that's fine," or a hug. Negative reinforcement is the withdrawal of an unwanted stimulus. Examples would include the cessation of a spanking, the disappearance of a parent's scowl, the removal of a pricking pin, the exit of an unpleasant person, or release from an unwanted responsibility, such as helping with the dishes. It is important to understand the distinction between negative reinforcement and punishment. Punishment also has two forms. Both forms discourage the repetition of a response. One type of punishment involves the presentation of an unwanted stimulus contingent on a response. An example would be the presentation of pain. Another would be social disapproval in its many forms. Still another would be the requirement of the performance of an unpleasant task. The second type of punishment is the withdrawal of a desired stimulus, such as prohibiting certain types of play activities, "grounding" the child, or not allowing the child to eat dessert. Both types of punishment diminish the probability of recurrence of the response that preceded them.

Primary and Secondary Reinforcers. Primary reinforcers are stimuli that feel, taste, or smell good, or are in some other way innately pleasurable, such as a push from a parent to a child in a swing, a piece of candy or stick of gum, a warm smile, a "that's good!", a movie, an affectionate squeeze. Secondary reinforcers are stimuli that can be redeemed or converted into something that is innately pleasurable. Poker chips, money, check marks on a chart, or tokens of any kind are examples. Secondary reinforcers are often preferable to primary reinforcers during therapy for the following reasons:

1. They can be presented or withdrawn, according to the response of the patient. To take candy away from a child when the child makes an inappropriate response is to arouse the child's emotions negatively toward the clinician and the therapy; to matter-of-factly take away a token is not so painful to the child.

2. The child does not have to be hungry at the moment to want to work for a token.

3. With primary reinforcers there is always the problem of whether to give the food, for example, to the child immediately or to wait until the end of the session. Either solution produces distractions. Tokens can be given immediately after correct responses.

4. Tokens are more flexible. A certain number of tokens can represent anything the clinician and patient, or parent and child, agree they will represent.

5. A hard-working patient can be "full" before the session is over if

primary reinforcers such as candy are used. This problem does not occur when tokens are presented.

Reinforcement Schedules. A child may be reinforced each time s/he produces a given behavior, or s/he may be reinforced inconsistently. In therapy, the schedule of reinforcement used initially is often "crf" (continuous reinforcement). Each time a child produces a desired response s/he receives the token, plus mark, or favorable comment from the clinician. The schedule may then progress to a fixed-interval or a fixed-ratio schedule. If the child is rewarded for every second correct response or every third or fourth one, the therapist is following a fixed-interval schedule. If the presentation of the reward depends on a predetermined ratio of correct to incorrect responses, the schedule being followed is a fixed-ratio type. The next schedule to be employed as therapy progresses is often the variable one in which the stimulus is presented in random, unpredictable patterns so that the child never knows when a given response is going to be rewarded. This type of progressive change in schedule has the advantage of gradually weaning the child from those types of reinforcement schedules that s/he is not likely to find in everyday, outside-the-clinic living. The child who is being assisted in the elimination of a thumbsucking habit must learn to keep the thumb out of the mouth even when there is no one else in the room. The tongue thruster must learn to swallow correctly under all types of conditions, day and night, and correct swallows are not going to be rewarded by a nod of approval after therapy has terminated.

The same principle, that of preparing the child for the reinforcement situations of everyday life by progressively making reinforcement more closely approximate what the child will expect to find in daily life, should be applied in determining the type of reinforcement to be utilized. If it is necessary, in order to achieve satisfactory motivation, to use primary reinforcers such as candy or alphabet cereal at the beginning of therapy, it is wise to structure the training to include a gradual elimination of these types of reinforcers. As therapy progresses, the candy should gradually be replaced by social reinforcers. These reinforcers should then, in turn, be presented with less and less frequency, until finally the child learns to depend totally on their own awareness of their correct swallowing patterns for reinforcement. That type of reinforcement is best, in other words, which most closely approximates the types of reinforcement the child will receive in everyday life.

Steps in Behavior Modification

Three basic steps are involved in operant conditioning: (1) establishing base line, (2) modifying behavior, and (3) extending stimulus control.

Establishing Base Line. "Base line" is the starting point for therapy. It is the behavior which the patient has in his repertoire that most closely resembles the behavior the therapist wants him to habituate. A given behavioral pattern, such as a tongue-thrust swallow, may require the establishment of several related base lines, due to the complexity of the muscle interaction involved. Base line for children who suck their thumb might be their behavior when they are in the presence of someone who is not a member of their immediate family. They may not suck their thumb in that situation but do so in all other situations, that is, when they are alone or with members of their family. The desired behavior is the cessation of thumb sucking. The behavior of the child with nonmembers of the family is the behavior that most closely resembles the goal behavior—the complete elimination of thumb sucking. It is a place from which to begin.

Modifying Behavior. Modification of behavior is often not too difficult while the patient is in the therapy situation. The behavior is "shaped" in a systematic, step-by-step manner until it conforms to the desired pattern. A shaping procedure that brings about a series of alterations in behavior, each of which brings the person nearer the ultimate goal, is called successive approximation. During this process, the appropriate reinforcers and/or punishments are used. Each step must be satisfactorily completed before the patient advances to the next step. The criteria for successfully completing each step are well defined and often expressed in terms of percentages. For example, a therapist may be helping a child to eliminate the dentalization of the linguoalveolar sounds /t/, /d/, /n/, and /l/. The child may be asked to tell a story that lasts for 5 minutes. The therapist counts the number of linguoalveolar sounds produced during the time period, and then tallies the number the child produced defectively. If more than 10% were defective, the child is not allowed to proceed to the next step, which may in this case be a part of the third step, extension of stimulus control.

Extending Stimulus Control. It is not difficult, in most cases, for a child to understand how to swallow correctly. Basically, the child is to form a seal between the tongue and the roof of the mouth, rather than between the tongue and the cheeks, lips, or teeth. The explaining can be accomplished in a relatively short time, but understanding is, of course, not the goal of therapy. The goal is to help the child make new muscle patterns so habitual that they become a part of their subconscious motor activity. This is the purpose of the third step, extension of stimulus control. We used to call this "carry-over" in speech therapy, back in the days when we called speech therapy "speech therapy." "Extension of stimulus control" is a better term because it more accurately describes the procedure involved in reaching the goal of subconscious habituation. In our opinion this step should not be considered accomplished until the myofunctional disorder has disappeared completely from the patient's behavior. This step must be equally as systematized as the preceding step. When therapy fails to be completely successful, it is most often due to a lackadaisical attitude on the part of the therapist, patient, or parent with regard to the specifics of achieving this step. This part of therapy is also the most misunderstood; many parents have the mistaken notion that once a child has learned to do something or has learned to stop doing something, failure to apply what the child has learned indicates laziness, lack of motivation, or negativeness on the child's part. A goal-directed plan is extremely important in this stage of therapy.

The therapy described in Chapter Eleven contains several applications of behavior modification principles.

Kinesiology (1976). We should take cognizance of some of the concepts presented under the general heading of kinesiology. Special applications of this field, specifically of a dental nature, have been appearing with regularity. Since kinesiology is customarily considered to be the study of muscles and muscular movement, it behooves the oral myologist to be aware of these developments.

Among the leading spokesman for this aspect of kinesiology have been Goodhart (1976), a chiropractor; Eversaul (1976), a psychologist; Diamond, a physician; and dentists such as May, Mittleman, and others. Fryman, an osteopathic physician who practices and teaches cranial osteopathy, might be considered a part of this movement.

While kinesiological procedures are often presented as an adjunct to orofacial myotherapy (or vice versa), they are sometimes proposed as the sole requirement for remediation. Suggested procedures, in addition to those designed to elevate the tongue and correct tongue thrust, include others to improve jaw relationships, to remove muscle spasm, to strengthen or weaken muscle reaction, to eliminate the gag reflex, to increase the range of mouth opening, and to change the fatigue rate in muscles throughout the body. For example, any muscle can be shown to fatigue more quickly if a few grains of refined sugar or other noxious substance is placed in the mouth. The same effect is shown even if the sugar, or a cigarette, is merely held in the hand rather than the mouth.

Another application of the foregoing is found in therapy localization. This is a diagnostic method in which the patient is asked to touch his fingertips to a suspected problem area, a TMJ, for example; should dysfunction be present, any muscle group in the arm or leg which has previously tested at normal strength will now show less resistance or will fatigue more quickly.

Remediation in basic kinesiology consists primarily of muscle manipulation. In cranial osteopathy it is achieved by adjusting the supposedly immovable bones of the skull.

There is also some reliance on "temple tapping." This is a procedure wherein the clinician offers a suggestion while tapping the patient's temple sharply and rapidly with fingertips; the suggestion is then rephrased while tapping the opposite temple. The effect of the suggestion or command is then manifested by the patient; for example, for a time thereafter the patient no longer gags when mechanically stimulated to do so.

Biofeedback (1976). Attempts to utilize biofeedback for the correction of oral myofunctional problems have been made in some quarters. When conventional equipment has been pressed into this service, the results have generally been disappointing.

Neuromuscular Facilitation (1977). Falk, Wells, and Toth (1976) developed a treatment program for tongue thrust based on procedures used by Margaret Rood. The approach is subcortical, designed to condition proper muscle tongue and action by stimulating the sensory receptors. Details of the method are presented by Falk (1977) in a later article.

Three types of stimulation constitute the treatment: brushing, icing, and pressure. The purpose of brushing is to reduce flaccidity of the tongue. The sides of the tongue are stroked repeatedly with a No. 6 oil paint brush, for ten seconds. The exercise results in a narrowing of the tongue and improved tonicity.

Icing consists of applying ice to the hard palate, in the area of the incisal papilla, for about ten seconds. This results in a reflexive elevation of the blade of the tongue to the area iced. Falk asserts that the focus of therapy should be to elevate the blade of the tongue, rather than its tip, as advocated by most clinicians.

Light pressure is applied to the dorsal and inferior surfaces of the tongue with the rounded end of a cocktail stick. The tongue is held in a relaxed position as the clinician taps the anterior part of the tongue in a rapid, alternating manner. The result is a posterior movement of the tongue according to Falk. Pressure to the lateral aspects of the tongue is applied in some patients to supplement the brushing exercise. Pressures are applied for only a few seconds.

Patients are instructed to carry out the above procedures before each meal and at least two other times in addition, for a total of five or six practice sessions a day. Sessions are brief, requiring only a minute or two. The regimen is typically continued for six months. Except for occasional attention to specific needs of individual patients, such as strengthening of the muscles of mastication, these procedures comprise the whole of Falk's treatment. He tested its effectiveness with an experimental group of eleven patients and found that all subjects experienced a reduction in severity of anterior malocclusions during the six months of therapy. Nine of the eleven demonstrated no regression during that period. There was no control group in this study.

In a subsequent study, however, Falk (1977) compared this type of treatment with a more traditional approach to tongue thrust. Two clinicians, both trained in procedures advocated by Daniel Garliner and the Falk et al., method, were each assigned two groups of five patients each, ages eight to twelve years. Each clinician treated five subjects with each approach. The subjects received six months of training.

Serial dental models were made before treatment, after three and six months of treatment, and six months following termination of therapy. Fourteen orthodontists judged the dental models in pairs, indicating whether one of the pair was "same," "better," or "worse" than the other. In addition, each judge indicated on a seven-point equal-appearing-intervals scale the amount of improvement each subject had made during the six-month course of treatment.

Both groups were judged as having made significant improvement toward normal occlusion following treatment. To a statistically significant degree:

1. Subjects treated with neuromuscular facilitation were more improved after three months of training than were subjects receiving treatment utilizing the Garliner approach.
2. A similar significant difference in improvement favoring the Falk treatment group was found after six months of treatment.
3. No significant regression in dentition was found in either group after six months following treatment.

Treatment Approaches

Current Approaches. We see three trends in approaches during the late 1980's: (1) A move toward preventive therapy. (2) Use of audiovisual aids (video, audiotapes) in evaluation and treatment. (3) In South America, a more holistic approach to orofacial myology, involving postural aspects of the spinal column.

1. **Preventive Therapy.** Three approaches are described in the literature. One says to begin taking preventive measures at the birth of the infant. A second attributes a great deal of anterior malocclusion and accompanying tongue thrust to digit-sucking. A third advised early work on tongue and lip resting postures.

Early Preventive Measures. Edger (1985) warns that neonatal feeding procedures may lead to "myofunctional imbalance." "Any object placed in this oral cavity on which an infant sucks, other than the natural shape of the mother's breast, will act as an orthodontic appliance, and will not only alter the correct resting position of the tongue, but will also create different potentials for malocclusion." (p. 23) Irregular pressures by the foreign object against the teeth may cause them to be intruded, crowded, or protruded. She advises breast feeding and avoidance of all unnatural-shaped pacifiers, nipples, and objects, particularly after 24 months of age.

Edger states that breast milk contains antibodies that protect the infant from upper respiratory tract infections, thus permitting establishment of predominantly nasal breathing patterns. She lists several signals of nasal passage blockage: bluish-black discolorations and bagging under the lower eyelids; dry, bulbous lips; constant tearing in the eyes; poor lip closure; high, arched palate; and nasal polyps.

Certain postural positions, continues Edger, foster anterior tongue resting postures, such as a forward-resting position of the child's head. "This forward head posture perpetuates a hyperactive state in the supra-hyoid muscle group, causing the tongue to leave its resting position." (p. 24)

A Nuk-type pacifier is recommended by Edger, along with the elimination or curtailing of gum-chewing and object-sucking.

Eliminating Digit-Sucking. VanNorman's (1985) practice in orofacial myology consists largely of children with digit-sucking habits. She attributes many of their anterior malocclusion to prolonged digit-sucking. "In my practice I have observed that 50 percent or more of those children with digit-sucking habits also have one or more teeth in posterior lingual crossbite." Varying degrees and types of open bite, she contends, are associated with digit-sucking and, virtually always, tongue-thrusting as well. She cites several studies to substantiate her claims, including the Hanson and Cohen (1973) research, that found digit-sucking to be positively correlated with a narrower, higher, and longer palate, and with shorter arch circumference, mouth breathing, dentalized speech sounds, and mentalis activity during swallowing. That research found digit-sucking to correlate negatively with amount of buccal crossbite and overbite. Digit-sucking was one of seven factors associated with the retention of tongue thrust patterns through the mixed dentition period.

VanNorman's approach to elimination of thumb- and finger-sucking uses positive reinforcement. She instills confidence in the child's mind about the outcome of therapy, and provides a system of reminders and reinforcers. She reports consistent success with the therapy, and advocates its use as a deterrent to the perpetuation of tongue thrust and to the development of anterior and lateral malocclusions.

Correcting Lip and Tongue Rest Postures. Pierce (1986) believes that a good portion of the success of therapy for tongue thrust is due to attention given to resting postures of tongue and lips. She advocates rest posture therapy for children five to nine years old. "Within (the seven to nine year old) group, there is a good possibility that a tongue thrust swallow will self-correct, perhaps with no therapeutic intervention at all. An abbreviated rest posture therapy program should enhance the changes that the swallowing pattern will progress through this transition."

She supplements work with rest postures with whatever tongue/lip exercises may be necessary to achieve those postures. Since the therapy is pretty much geared to younger children, Pierce's exercises and activities are kept simple. Reminders of various kinds, such as stickers and signals, are used. "To help you remember just where your SPOT is, Mom can cover it with toothpaste, honey, vanilla extract, or other goodies."

Some examples:

Pretend your lower lip is a balloon. Blow up the "balloon" and hold for five seconds.

Fat Tongue/Skinny Tongue. Make your tongue skinny for three seconds, then make it fat for three seconds.

M & M Squash. Keep and M & M against the "spot" with your tongue tip. HOLD it there till the candy coating breaks.

Positive reinforcement techniques are used to motivate the child.

2. **Use of Audiovisual Aids.** Bill and Julie Zickefoose (1988) have been leaders in the utilization of audio and video recordings in the evaluation and treatment of orofacial pattern disorders. At the authors' request, The Zickefooses provided a summary of their rationale and procedures. Following are excerpts from that summary.

Video Recording. The majority of patients and parents who enter the myofunctional therapist's office for the first time have little knowledge of myofunctional disorders. Therefore, it is highly important for the therapist to be able to give and show them as much information about myofunctional disorders as possible. In most cases the therapist has little equipment in the office for diagnostic purposes. The standard procedure has been to have a still camera and show the patient and parents slides or photographs that show various types of myofunctional disorders; however, slides and photographs cannot show *function.*

In myofunctional therapy, "function" is the key word. Video tapes will show function and should become a part of a therapist's program. Because the therapy program is continued over a long period of time, it is very difficult for the patient, parent, and even the therapist to remember how severe the myofunctional disorder was at the beginning of therapy. Through the use of video equipment, one can show progress, thus increasing motivation for both patient and parent. Without this type of motivation it is often hard to keep the patient interested in the MFT program until habituation can occur.

Video Taping Procedures. Make sure the camera is level with the patient's head and the lighting is adequate. For the best video quality, a dark blue background is recommended. Ask the patient or patient's guardian for permission to make and use the recording. Follow the following outline of procedures.

1. Evaluate the patient's speech patterns.
   a. Count to twenty.
   b. Read or repeat a short list of words.
   c. Participate in a short conversation.
2. Examine oral structures and functions.
   a. Check the condition of the hard and soft palates.

b. Assess size and movements of the tongue inside and outside the mouth. Have the patient pop the tongue.

c. Use cheek retractors to permit examination of the dentition.
1. Record lateral and frontal views of the teeth.
2. Tilt the head back to show overjet, if present.

d. Using a transparent cup, check drinking.
1. Observe contraction of circumoral muscles.
2. Part lips as the swallow begins to check for tongue fronting.

e. Use two consistencies of food to check eating behaviors.
1. Check right lateral and frontal views while the patient eats a banana or an apple.
2. Check left lateral and frontal views during the eating of a cracker or wafer.
3. Video the patient's sitting and standing postures. Notice any forward posturing of the head, slumping or uneven shoulders, or curving of the spine.
4. Close the video session with frontal and lateral views of the patient's face. If the patient, according to your prior instructions, has brought close-up photographs taken at earlier ages for comparison, place the lens adjustment on macro and state the patient's age as you record each photo.
5. Review the tape with patient and parents. Point out significant characteristics, such as facial grimaces, a short lingual frenum, a glide of the mandible, a high narrow palate, or whatever myofunctional disorders you might have discovered.
6. Inform patient and parents that the taping procedure will be repeated in a few months.
   If at any time the therapist observes noticeable improvement, the patient should be video taped for motivational purposes. Subsequent recordings are added to the original tape. Each time a new tape is made, the preceding tapes are reviewed so that comparisons may be made.

Tape Recording the Lessons. When lessons are tape recorded, rather than present to the patient in written form, exact instructions are given orally and listened to at home. Each exercise and the specific number of repetitions required are recorded. When the taped lesson is completed during each practice session, the patient may go about the rest of his/her daily routine.

Recording the lesson is a very simple procedure. Determine the type of exercise program that would be appropriate for your patient. For each exercise, repeat the exact number of times you wish the patient to practice. During the recording the therapist makes suggestions, such as, "Watch for facial grimaces," "bite your back teeth together," "feel the lift of the back of your tongue," or "I really like the way you are watching your mirror." This provides constant reinforcement of what the patient is to look for and feel, along with words of encouragement.

Usually the lessons take only from five to eight minutes. The patient is asked to practice this tape three times each day. Actual practice time is much shorter than with a written lesson, because practice is much more concentrated and guided.

With the session tape recorded, it is easy for the therapist to assess the patient's practice. When the patient returns for the next visit, place the tape in your recorder and ask the patient to perform the lesson for you. If the patient anticipates the tape sequence and performs well, you can be sure that a sufficient number of practices have been completed.

3. **Attention to Posture and Flexibility of the Spinal Column.** Beatriz Saboya, owner of a large private speech and hearing clinic in Copacabana, Brazil, presented a paper in 1983 at a phonoaudiological congress in Rio de Janeiro, and later published the paper in the *IJOM* (1985), in which she urged that greater attention be paid, in evaluation and in treatment, to the alignment of the spinal column in patients with oral myofunctional problems. She criticized those who take a "fragmented" approach to tongue thrust by treating only the mouth. She defines the "axis" as the longitudinal line that divides the human body in approximately symmetrical or balanced parts and the line around which a human makes rotational movements. "In our statistics, about 60% of the patients with atypical swallowing have an irregularity of the axis. We note that patients who have the head habitually inclined forward and downward demonstrate a *tendency* toward a class III and those who have the head fixed more toward the back tend toward a class II." And further, "When we pay attention only to oromyofunctional abilities and to respiration, therapy does not produce the same results that we get now, when we include work with the axis at the beginning of the program of rehabilitation, given the existence of an irregularity." Therapy time is shortened, she affirms, when therapy attends to the postural and movement aspects of the spinal column.

The evaluation includes static and dynamic characteristics of the axis. The patient is observed standing and sitting, from various vantage points. Any asymmetries are noted, as well as irregularities in the curvature of the spinal column. Then the patient is viewed while running, walking, and jumping with feet together. Indications of disharmonies of parts, or of inflexibility, are noted. Therapy is planned according to these findings. Often the speech therapist works with the physical therapist, to improve general posture and flexibility of movement. Once these conditions are corrected, attention is directed toward the more conventional aspects of therapy for oral myofunctional disorders. The approach seems worthy of our study and consideration.

Summary

An orthodontist faced with a patient with a significant tongue thrust problem has several choices: Ignore it and hope it goes away; offer a simple suggestion to the patient that s/he keep the tongue against the roof of the mouth and away from the front teeth; refer for oral myofunctional therapy; place some kind of tongue-repelling appliance in the mouth; refer for tongue-reduction surgery; or combinations of these choices. While each approach has a few adherents, most orthodontists who recognize the problem make referrals for therapy. They recognize the limitations of appliances for most patients, and the futility of a simple suggestion in attempting to change complex behavioral patterns. In a small number of cases, where motivation is not achievable through therapy, or when, for any number of other reasons, therapy fails, a reminder appliance may have to be used.

If therapy is chosen, there are a number of approaches to consider. The chapter has reviewed salient characteristics of a number of them. We recommend that as you evaluate any program you look for answers to the following questions.

1. Does every exercise and assignment have a specific purpose?
2. Is there any unnecessary duplication of exercises?
3. Is the succession of skills planned so that each new step builds on previous ones?
4. Is it adaptable to the needs of patients of all ages?
5. Does it allow for periodic reevaluations of the patient's progress?
6. Does it include proper emphasis on resting postures of orofacial muscles?
7. Are appropriate motivational procedures included?
8. Does it provide for necessary feedback to the patient, his/her parents,

and the therapist, so that each can objectively know of the patient's progress?

9. Does it include attention to all related orofacial habits?

After choosing a program to follow, learn it thoroughly, by hearing it described by one who knows it well, then studying it carefully, observing it being administered to a number of patients, then trying it out, with frequent observations and feedback from a competent user of the program. Dare to deviate from that program, but always base changes on sound rationales. Finally, individualize the program for each patient, according to his/her needs, progress, and difficulties with its steps.

REFERENCES

Barrett, R. H., and von Dedenroth, T. E. A. (1967). Problems of deglutition. *Am. J. Clin. Hypno., 9,* 161.

Cleall, J. F. (1965). Deglutition: a study of form and function. *Am. J. Orthod., 51,* 566.

Crowder, H. M. (1965). Hypnosis in the control of tongue thrust swallowing habit patterns. *Am. J. Clin. Hypno., 8,* 10.

Edger, R. (1985). Early intervention in myofunctional imbalance. *Intern. J. Oral Myology, 11,* 22–25.

Eversaul, G. A. (1976). Biofeedback and kinesiology: technologies for preventive dentistry. *J. Am. Soc. Prevent. Dent., 6,* 19–23.

Falk, M. I., Delaney, J. R., and Litt, R. I. (1976) *Comparison of selected cortical-level and reflexive-level treatment programs for establishing normal deglutition patterns.* Unpublished paper.

Falk, M. L., Well, M., and Toth, S. (1976). A subcortical approach to swallow pattern therapy. *Am. J. Orthod., 70,* 419–422.

Falk, M. L. (1977). Treatment of deviant swallow patterns with neuromuscular facilitation. *Int. J. Oral Myology, 3,* 27–29.

Goda, S. (1968). The role of the speech pathologist in the correction of tongue thrust. *Am. J. Orthod., 54,* 852.

Goodhart, G. J. (1976). Kinesiology and dentistry. *J. Am. Soc. Prevent. Dent., 6,* 16–18.

Graber, T. M., and Neuman, B. (1977). *Removable orthodontic appliances.* W. B. Saunders.

Kaye, S. R. (1973). Modified appliance to control tongue-thrust. *Dental Survey,* 24–25.

Lambert, C. G. (1959). Hypnotherapy in control of tongue thrust occurring during perverted swallowing. *Am. J. Orthod., 45,* 869.

Lines, P. A., and Steinhauser, E. G. (1974). Diagnosis and treatment planning in surgical orthodontic therapy. *Am. J. Orthod., 66,* 378.

Littman, J. Y. (1968). A practical approach to the tongue-thrust problem. *J. Pract. Orthod., 2,* 138.

McWilliams, R. R., and Kent, J. N. (1973). Effect on protrusive tongue force of detachment of the genioglossus muscle. *J. Am. Dent. Assoc., 86,* 1310.

Palmer, G. L. (1961). The effectiveness of post-hypnotic suggestion upon control of the tongue-thrust habit in relation to trance depth. *Am. Orthod.*, *457*, 228.

Pierce, R. B., and Warvi, V. (1973). *Swallow right.* Huntsville, AL: Huntsville Rehabilitation Center.

Pierce, R. (1986). Rest posture therapy. *Intern. J. Orofacial Myology*, *12*, 4–12.

Saboya, B. (1985). The importance of the axis in the study of oromyofunctional disorders—integrated approach. *Intern. J. Orofacial Myology*, *11*, 5–13.

Secter, L. L. (1961). Tongue thrust and nail biting simultaneously treated during hypnosis: a case report. *Am. J. Clin. Hypno.*, *4*, 51.

Straub, W. J. (1962). Malfunction of the tongue, Part III. *Am. J. Orthod.*, *48*, 486.

Subtelny, J. D., and Sakuda, M. (1964). Open bite: diagnosis and treatment. *Am. J. Orthod.*, *50*, 337.

Tepper, H. W. (1986, December). Tongue thrust correction in one each lesson. *J. Amer. Acad. of Gnathologic Orthopedics*, *3*, 4–5.

VanNorman, R. (1985). Digit-sucking: it's time for an attitude adjustment or a rationale for the early elimination of digit-sucking habits through positive behavior modification. *Intern. J. Orofacial Myology*, *11*, 14–21.

Walker, R. V., and Collins, T. A. (1971). Surgery or orthodontics—a philosophy of approach. *Dent. Clin. North Am.*, *15*, 771.

Whitman, C. L. (1964). Correction of oral habits. *Dent. Clin. North Am.*, 541.

Zickefoose, W. E., and Zickefoose, J. (1988). Personal communications.

CHAPTER 10

A PHILOSOPHY OF TREATMENT

The most important elements in any therapy program are the philosophy on which it is based and the attitude of the therapist that is reflected by that philosophy. It may develop that the reader will wish to modify many of the procedures that will be presented in the balance of this volume; if you know why you are changing it, and how this change relates to the ultimate goal, and if you make the adjustments with the proper attitude, you can only be successful. The specific techniques that will be set forth are merely those with which the writers feel comfortable; they may adapt poorly to the personality of a different clinician. They are meant to achieve a certain purpose; it is the goal that is important, and these techniques will be some of the possible courses by which it may be reached. They in no way exclude alternate routes.

Nine statements express the basic elements of our philosophy of treatment:

1. The approach should be psychophysiological.
2. The purpose of therapy is to effect changes in behavior that reach an automatic level.
3. Therapy must be a cooperative endeavor among the patients and all who deal with them, including parents or spouse, dental specialists, medical specialists, teachers, and therapists.
4. Therapy procedures should incorporate all that is known from research and from clinical experience.
5. Treatment should be preventive whenever possible.
6. Some skills to be taught are more basic than others.
7. The therapeutic process cannot be shortcutted without impairing its effectiveness.
8. A number of factors must be weighed in making decisions concerning treatment timing.
9. Therapy for tongue thrust is usually successful.

1. The Approach Should be Psychophysiological. It should be individualized, and should take into account everything the therapist can learn about the patient from those who live with and deal with him/her. Critical past, present, and future stimuli should be considered in the planning and administering of therapy. Three basic principles are important:

a. It is not possible to "break" a habit; accordingly, we must replace one habit with another habit. Although therapy replaces unwanted psychomotor patterns with new ones, the former are always subject to recall under certain physiological and emotional conditions. Parents, patient, and clinician should always be alert for any signs of relapse.

b. Attention to each of the component parts of a behavioral pattern is necessary when that pattern is an integration of several related components. This is true in the case of orofacial myofunctional disorders. Therapy must encompass the total problem and all its aspects.

c. Habits gain strength as time passes. Undesirable patterns should be eradicated as soon as is feasible, and new ones should be established as soon as the patient is mature enough to carry out necessary assignments.

The most successful therapist combines knowledge of techniques for changing behaviors with an understanding of human nature and the ability to *motivate.* The clinician must convey sincere interest and enthusiasm and arouse and maintain that same level of enthusiasm in the patient.

2. The Purpose of Therapy is to Effect Changes in Behavior That Reach an Automatic Level. This principle and the first principle are strongly related. The application of principles of operant conditioning can change behaviors, but those principles must be accompanied by solid interpersonal skills, particularly at the stage of extension of stimulus and response controls. Old patterns, usually of several years' standing, resist total extinction, and newly learned patterns must be repeated hundreds of times before they become automatic.

A brief review of behavior modification principles seems warranted here. The three basic steps are:

a. **Establish Base Line.** This step consists of determining precisely what the patient is doing correctly and what behaviors need to be changed. During the consultation, a complete case history is obtained and behavioral patterns are observed and recorded. Direct questions concerning motivation and attitude are often productive in establishing base line. The clinician observes behavioral patterns carefully. If tongue thrust is present, what form does it take? Does the thrusting occur during the swallowing of saliva, liquids, and foods? The strengths of involved muscles and the appropriateness of the timing and nature of their movements during function are assessed. A patient may not habitually occlude the molars during swallowing, but the masseter and temporalis muscles may still be normal in strength. If that is the case, the patient would be assigned exercises to increase *awareness* of the contraction of those muscles, and to systematically establish the habit of their contraction during swallowing, rather than to work to strengthen them. Does the tongue rest habitually against the upper or lower anterior teeth, or between the maxillary and mandibular teeth? Is there a lisp? How stimulable is the patient when given a brief explanation of how to swallow correctly? Might the base line be altered if remediable conditions, such as large tonsils or adenoids or a deviated septum, were treated medically or surgically?

Another important procedure in the establishment of base line consists of having the patient and family members observe during the week following the consultation various aspects of the habits needing correction. If the problem is digit sucking, they are asked to observe under what conditions the finger or thumb is placed in the mouth, the manner in which it is positioned, the length of time it is left in place, and whether there is any relationship between the occurrence of the sucking and fatigue, emotional state, time of day, hunger, or happenings during the day. In the case of tongue thrusting, observe when the mouth is most likely to be held in an open position; where the tongue usually rests when the mouth is open; the relative involvement of tongue and lips during the chewing process; the positions of the tongue and lips during sleep; and any other abnormal behaviors involving the oral musculature.

The description of base line assures the therapist of seeing the total pattern and provides for individualization of therapy. It eliminates unnecessary and ineffectual use of stock exercises and assignments and greatly improves the quality of therapy.

b. Modify Behavior. This phase of therapy can be completed for all but the youngest patients with the use of social praise as the only reinforcer. For younger children, and for older children who are difficult to motivate, token or primary reinforcers are used. Whenever possible, children are made responsible for their own practice and for keeping current the practice charts they are to bring to each therapy session. If for any reason the parent is unable to observe a given practice session and wants to know whether the practice has been accomplished, s/he need only look at the practice chart. For the patient, the assurance of being trusted to practice and keep up the practice chart provides good motivation.

Throughout this phase of therapy, certain subgoals are set that are related to the achievement of skills involved in the swallowing process. These are presented to the child in list form at the beginning of therapy. With the younger child the parent and child are asked to decide on a small reward for the completion of each of the subgoals. Reaching a major goal brings a greater reward. The following are a few achievements that might be included on the list:

1. Tight seal between blade of tongue and roof of mouth.
2. Ability to chew while keeping lips closed and tongue away from anterior teeth.
3. Posterior movement of saliva without pushing tongue against teeth.
4. Drinking of liquids without unnecessary movement of facial muscles.

It is important during the behavior modifying stage of therapy for the clinician to define very precisely the role of the parent in providing reinforcement or punishment. Probably most important is the insistence to the parent that all types of punishment, other than that prescribed by the therapist, be avoided. It is also important to avoid using facial expressions that convey disgust or anger to the child who misses practices or performs an exercise in an imperfect manner. Appropriate praise and encouragement are usually more effective than punishment.

   c. **Extend the Stimulus Control.** This is the "carry-over" part of therapy. The patient has learned proper lip and tongue posturing and is able to handle foods, liquids, and saliva correctly. Now all corrected patterns must be habituated. In truth, carry-over training begins when patterns are being learned, but now structured emphasis is given this phase. Whatever type of reinforcement has been used during therapy has to be gradually modified to conform with the very sporadic and infrequent feedback that will occur in everyday life. Parents, siblings, friends, and teachers should be called upon to help, unless the child objects to assistance from any of these people. The feedback is less objectionable if it is given in the form of a simple signal, such as touching the forefinger to the thumb, or touching the cheek or chin to let the child know that the mouth is resting in an open position or that the tongue is resting against the teeth. The hundredth time the child receives the signal is less disturbing than the hundredth time the parents say to close the mouth. Children are provided with reminder buttons, stickers, signs, or posters to place in conspicuous places in the house and in or on the desk at school. They may be asked to make their own reminder signs, or to take bookmarkers to school with messages written on them. They may be given a list of reminders from which they are instructed to choose two or three, and these may change every week or two. Details of this stage will be provided in Chapter Eleven.

3. **Therapy Must be a Cooperative Endeavor.** Each party in the cooperative endeavor has an important role. If any fails to do his/her part adequately, the prognosis for success in therapy dims significantly.

a. The Referring Dentist. If the referral is not made, of course, therapy will not occur. Many referrals are made but the patient or parent does not contact the clinician, and the clinician may never know of the referral. A good remedy for this is for the clinician to provide triplicate referral forms for use of referring specialist. When the dentist or physician makes a referral s/he gives the patient a copy, keeps a copy, and sends a copy to the therapist. If the patient has not contacted the therapist within a reasonable period of time, the therapist contacts the patient. The therapist informs the referral source as to whether contact has been made and a consultation date set.

The manner in which the referral is made is extremely important. A few orthodontists seem to feel they can motivate patients to try hard in therapy by telling them how difficult therapy will be, or how unlikely will be success. Others make the referral hurriedly, with little explanation of its importance. The dentist who explains the necessity of the therapy, describes briefly what its effects are likely to be, and conveys a positive attitude toward it, will refer patients who actually make the appointment for the consultation for therapy and follow through with treatment.

It is helpful to the treatment planning of the therapist if the orthodontist who refers patients explains any special conditions surrounding the total treatment of the patient. If the palatal arch is to be expanded, if treatment is going to be two-stage, if a headgear is to be worn for a period of time preceding banding, if surgery is to be performed, or if orthodontic work has been done previously, the clinician should be so informed. When therapy reaches the carry-over phase, and the dentist notices signs of relapse in the patient, that should be communicated to the therapist, and the patient referred back for supplemental training.

b. The Therapist. The therapist needs to bring to the partnership the proper training, experience, and attitude to carry out the program efficaciously. Helpful characteristics include flexibility, dedication, imagination, a sense of humor, a love for people of all ages, patience, perceptiveness, a reasonable degree of intelligence, an ability to explain concepts and instructions clearly, and a measure of forgiveness.

The clinician needs to communicate the rules of therapy to patient and parent. We require three practice sessions daily, seven days each week. If any two practice sessions are missed during the week, we expect a phone call informing us, and we postpone the next session for another week. The child who understands this, and sees the rule reinforced appropriately, is much more inclined to practice consistently than the one who recognizes a looseness in the application of a nebulous practice rule. Repeated expressions of negativity or reluctance to practice by the child may result in discontinuation of therapy. The child is always welcome to return, when s/he makes a decision to comply with practice requirements.

Many patients encounter *difficulty* in some aspect of therapy; the program may require twice its planned duration before mastery is achieved. This is to be expected, and the clinician should be willing to make any allowance and provide extra assistance as indicated in support of a patient who is having trouble but honestly *trying* to make progress.

This demanding attitude on the part of the clinician has proven to be in the patient's best interest in our practices. When uncooperative patients have been continued in therapy week after week, the result invariably has been stagnation of attitude, nil progress, and eventual failure. On the other hand, a child dismissed from therapy for lack of effort has very rarely failed to return with a new appreciation of the importance of cooperation.

The fuel on which patients operate best is a blend of praise, kindness, and approbation. Even a mild display of effort should be stoked with this combustible compound. However, children have a marvelous ability to detect insincere or undeserved praise. A phony "sweetness and light" attitude by the therapist may produce a negative effect.

Praise need not be restricted to specific behaviors. Approbation of color of eyes, neatness of dress, or well-groomed hair does wonders for rapport. During oral examination, it is better to replace perfunctory grunts with some display of elation at even the most commonplace lingual ability—not that the average person could not do these things, but you are delighted to observe them in a tongue thruster. Let the child see at the outset that you are prepared to warmly reward the effort s/he expends and reflect the assumption that *of course* s/he will do well.

In the rare instances in which suspension from therapy becomes necessary, the attitude of the clinician will, of course reflect some regret and disappointment. We expected better things of the patient, we have demanded nothing that could not be done easily had the desire to do so been there, and we are infringing on only a very brief segment of the patient's life. If the necessary effort has not been made up to this point in therapy, the more difficult things farther ahead in treatment would not have been accomplished either. Nothing could please us more than a telephone call at any time in the future, wherein the patient expresses a desire to resume treatment. If the child returns, the clinician is visibly happy, but explains that the rules have not changed.

As is the case with salesmen who must make an oft-repeated pitch seem fresh and new to each client, the therapist must have the ability to convince each patient that s/he is special and different, and that the words the clinician utters are spontaneous and sincere. This is easier to do when it is really true.

It is beneficial occasionally for the therapist to "sit on the other side of the table," to mentally view therapy from the patient's perspective. One insight quickly gained is an appreciation for the patient's need to understand, to know what is going on. In order for a necessary new skill to be fully and efficiently incorporated into his/her neuromuscular system, the patient should clearly see its rationale. It is not enough for the therapist to merely assign exercises. Instead, the clinician must have the knowledge, ability, and the patience, to explain *why* the procedure is important, what, specifically, it is designed to accomplish, and how this will be integrated into the total outcome. In matters of orofacial myology, we have long subscribed to the belief that "therapy occurs five inches above the mouth." That is the therapist's assignment.

c. The Patient. Two types of attention are required of patients during therapy. First, they need to maintain a general awareness of tongue and lip resting postures through therapy. Second, they must be willing to practice, first exercises, and later in therapy, assignments in eating, drinking, saliva swallowing, and pattern strengthening. Practice time rarely exceeds a half hour a day, but awareness time has to be pervasive. Younger patients can be helped by parents with both types of attention, but patients nine years old and older should be responsible for their own compliance with assignments.

It is not feasible for young children to be motivated by the healthy, pretty teeth they will have as adults if they do good work in therapy. They do need to be motivatable in *some* manner, though, by some more immediate and tasty kind of reward, and they do need to possess a reasonable attention span. They need to relinquish control of the therapy session to the therapist. They need to agree to practice willingly, rather than reluctantly, for they will reach a point in therapy when urgings from parents are not enough to continue the progress. Sometimes they need to be understanding and forgiving of parents, who are usually new at the task of assistant therapist.

For some older children this therapy regime is their first experience at being entrusted with true responsibility and they welcome the opportunity. They must make a commitment to carry out all assignments without fudging, or to communicate their practice misses honestly. They must be willing to do a lot of mirror watching, particularly at meals. If they need to be reminded to practice, or if they are signaled frequently for not having their lips closed at rest, they must receive the reminding graciously. As noted above, they usually do better if the reasons for exercises are explained to them; they have a right to know them.

Adults are often the easiest, most rewarding patients of all. They may be skeptical about their ability to learn "new tricks," but are convinced when exercises make tongue movements easier and the new lingual resting posture actually feels comfortable and natural. They need frequent reassuring about the period of time required for habituation of new patterns. Progress is facilitated when adult patients are willing to receive signals and prompting from other persons.

d. Parents. Parents need to be informed of the essential nature of their role. Some tend to be too intrusive, others apathetic. Understanding the proper amounts of reminding, urging, and reinforcing they should do, they must be willing to give this therapy its due time and attention. For a period of several weeks, it must be placed high in the hierarchy of family activities. Canceled appointments never result in death or even serious illness; they thus occur frequently unless the parents appreciate the importance of continual progress to motivation of the child. We provide the parents, at the consultation session, a detailed explanation of what is expected of them.

We almost always have parents sit in on therapy sessions, unless either the parent or child requests otherwise. In this way the parent learns the purpose and nature of the exercises and assignments, and also learns what to look for should relapse occur. We make the parents responsible for monitoring the child's progress during all stages of therapy, and following the completion of the formal lessons.

4. Therapy Should be Based on Research and on Clinical Experience. Neither has exclusive rights to value. Clinical insights and experimentations always precede research investigations into their usefulness.

Some principles based on research findings are:

a. The most basic patterns to correct are lip and tongue resting postures.
b. Unless nose-breathing clearly predominates over mouth-breathing, the prognosis for success of treatment is poor.
c. Generally, muscle *movement* retraining is more important than muscle strengthening. Research has shown that therapy with minimal, or no, strengthening exercises is effective in changing tongue and lip patterns and in reducing orthodontic relapse.

Some principles based on clinical experience are:

a. Although abnormal lip and tongue resting postures, and tongue thrust during chewing and swallowing of food, involve pressures that are more harmful to teeth, tongue pressures that occur during drinking, saliva handling, and speech are a part of a total integrated pattern and should also be eliminated if therapy is to result in permanent pattern changes.
b. Many principles of good speech therapy apply to orofacial myofunctional therapy, such as individualizing treatment, keeping adequate records, cooperating with other professionals, and involving parents.
c. The therapist should become affiliated with one or more professional organizations, take advantage of workshops in orofacial myology and related areas. This prevents becoming stale in one's treatment procedures.
d. Every exercise and assignment should have a definite purpose, one that should relate in proper sequence with other therapy activities and with long-term goals of treatment.

5. **Treatment Should be Preventive Whenever Possible.** Years ago we recommended not seeing patients until the age of seven. There were two basic reasons: (1) Younger patients lacked the maturity necessary to carry out assignments. (2) Many children younger than seven were still likely to self-correct, spontaneously. The emphasis on preventive treatments of all kinds has led us to alter this opinion. We found that many children as young as four years of age are able to benefit from at least partial treatment, enough to prevent developing malocclusions from worsening. If the child is reasonably intelligent and mature, the parents are concerned and cooperative, and if there is evidence of habits that are

contributing to adverse movements of teeth, we attempt treatment. Most children are able to learn to keep the tongue in the proper place at rest, and to keep the lips closed. Most can learn to chew with the lips closed and to swallow food and liquids without a tongue thrust. With considerable help from parents, most are able to maintain corrected patterns. Frequently these young children are not able to habituate proper collecting and swallowing of saliva, but we have seen enough exceptions to this that we often even work on saliva handling.

Those children who have difficulty breathing through the nose at this age are referred to an otolaryngologist. Most are helped to a significant degree, through medication or surgery. This early preventive work permits normalization of resting postures of lips and tongue, important to the soon-erupting permanent teeth.

6. Some Skills are More Basic Than Others. In *descending* order of importance, ranked on bases of research and clinical experience:

a. Resting Postures of the Tongue and Lips. These constitute the beginning position for all muscle movements related to tongue thrust. The law of physiologic economy applies here: If the tongue rests against the anterior teeth habitually, and returns to that position after any movements, such as those occurring in chewing, swallowing, or speech, it will probably move away from that anterior position as little as possible during the activity. Thus, resting postures become the key to the frequency and extent of linguodental contacts during the day and night. Orthodontists and researchers agree almost unanimously on the greater importance of light, constant forces, such as are found in resting postures, over the stronger, less frequent lingual pressures against the teeth during swallowing. For these reasons lingual and labial resting positions receive primary emphasis in our program.

b. Saliva Moving. Most programs give inadequate attention to the preparatory phase of saliva handling. Saliva normally collects forward in the mouth, under the tongue. It must be moved posteriorly without a tongue thrust. If the saliva swallow is corrected, but saliva moving continues to be effected by a linguodental seal, the tongue thrust is likely to return during swallowing of saliva as the weeks and months go by.

c. Saliva Swallowing. Saliva handling is given precedence over food because of the frequency of saliva swallows and because of the strong relationship between saliva handling and resting postures of the lips and tongue.

d. Food Chewing. Chewing pressures may occur from five to twenty times per mouthful of food. Proper chewing involves very little contact between the tongue and the anterior teeth. The patient is taught to allow the lips to move the food posteriorly, and the tongue to move it laterally.

e. Food Swallowing. The efficiency of the food swallow is stressed. If food remains in the mouth after an initial swallow, an additional swallow is required, which is nearly always preceded by a "clean-up" action of the tongue against the front teeth. The tongue thrust swallow may have accomplished the same task with only one swallow, and again the law of physiologic economy will promote a return to the more efficient type of swallow. The preparatory phase of swallowing, after chewing is completed, is carefully explained to the patient.

f. Speech. Speech pathologists usually notice dentalized /s/ and /z/ sounds, because the anteriorization changes the acoustical characteristics of these sibilants. Dentalization of other linguoalveolar sounds, such as the /t/, /d/, /n/, and /l/, however, may escape detection. These latter sounds involve much greater pressure against the front teeth than do the /s/ and /z/, and should be corrected by a speech-language pathologist. If the orafacial myologist does not have the necessary training to be able to work on speech, s/he should make a referral. Our approach is to work on dentalization as a distinctive feature, rather than to correct the problem one sound at a time.

g. Liquids. Liquid swallows occur relatively infrequently, and require very little pressure. Since they are often a part of the total problem, nevertheless, they should be corrected. Therapy includes instruction in continuous drinking, sip-at-a-time drinking, and drinking from a fountain and through a straw.

7. The Therapeutic Process Cannot Be Shortcutted Without Impairing its Effectiveness. Straub's original program for correcting tongue thrust contained 21 lessons, each filled with a variety of exercises, including several for speech. There are still therapists who see patients for as many as 30 visits. Most programs, though, require from seven to 15 visits, followed by a variable numbers of rechecks. A probable mean number of times patients with tongue thrust see their therapists in the United States would be 15.

The ultimate in shortcutting therapy is the well-intentioned dentist who shows the patient where the tongue should rest, tells him/her to keep it there always, including during swallowing, and moves on to the next patient, satisfied the job has been done. Therapy has to deactivate some muscles, activate others, integrate muscle activity into carefully sequenced patterns, and make those new patterns natural and consistent.

That process requires considerable structured home practice. The dentist or therapist who attempts to over-simplify the process is not doing the patient a favor.

8. A Number of Factors Must be Weighed in Making Decisions Concerning Treatment Timing. Two considerations are basic: (1) the age of the patient, and (2) treatment by other specialists.

Age of Patient. We have discussed the value of preventive treatment for children as young as four years of age. Children ages six to nine who are passing through the mixed dentition state, vary a great deal in terms of maturity and attitude. Those children who have significant malocclusions or speech disorders, and who appear to have enough maturity and parental cooperation, and who are motivatable to practice, do well in therapy.

Most patients referred to orofacial myologists are in the 10 to 17 year old group. Patients in this group are usually highly motivatable. A few are rebellious or simply too busy to bother with the therapy regimen. The success rate with this age group is very high, in general.

Many referring orthodontists doubt the ability of adults to respond to therapy. In our experience, adults are the best prospects for success in treatment. They often question their ability to change such deeply ingrained habits but are convinced after a couple of weeks of doing exercises that habits are indeed changeable.

Treatment by Other Specialists. Aside from age of the patient, another important consideration in treatment timing is treatment that must be done by other specialists. Should therapy generally precede, accompany, or follow orthodontic treatment? There are some advantages to correcting the habits before the orthodontic work begins. The patient who has been told by the orthodontist that the initiation and success of orthodontic treatment depend on the successful completion of the therapy program begins therapy with an important motivational advantage. When therapy is successful, the orthodontist, the family of the patient, and the patient have considerably more assurance that results of orthodontic treatment will be permanent.

Those who prefer to delay therapy until orthodontic treatment is completed argue that many times the orthodontic treatment will correct the tongue thrust without any therapy. Research done by Andrianopoulos and Hanson (1987) found this to be not true. Some therapists feel that the orthodontist provides the patient with a more normal environment in which the tongue may function. While this may be true, in most cases tongue thrust therapy is very effective even though malocclusions may be present. Exceptions are severe anterior overjets and openbites. If possible, partial correction of the malocclusion prior to the beginning of training of lip and tongue resting postures is helpful when malocclusions are severe. In our experience, however, orthodontists have preferred to delay orthodontic treatment until some therapy has been provided.

Our preference is to see the patient for four or five months before orthodontic work begins, and to follow the patient until all orthodontic treatment is completed. There are some exceptions to this procedure. If the palatal arch is extremely narrow and expansion is necessary, it is helpful to have this expansion done before therapy begins. If the patient is unusually young or immature or is not capable of doing well in therapy, and if preventive orthodontic work seems advisable, it is sometimes necessary to delay the therapy accordingly.

9. **Therapy for Tongue Thrust is Usually Effective.** The preponderance of evidence in the literature finds therapy for tongue thrust to be effective and permanent. Of 14 studies, 13 report favorable results. These are Robson (1963), Barrett and von Dedenroth (1967), Stansell (1969), Christofferson (1970), Case (1975), Overstake (1975), Toronto (1975), Cooper (1977), Christensen and Hanson (1981), Ohno (1981), Harden and Rydell (1983), Young and Vogel (1983), and Andrianopoulos and Hanson (1987). Therapy was reported to be ineffectual by only one writer, Subtelny (1970), who studied five subjects with "abnormal swallows." Only three of the five subjects demonstrated true tongue thrust.

The most recent of the studies (Andrianopoulos and Hanson, (1987) found therapy for tongue thrust to be effective in limiting orthodontic relapse. Thirty four subjects, ages 16 to 30 years, all of whom had worn braces and an upper arch retainer, and all of whom had been out of retention for at least one year, were examined. The seventeen experimental subjects had received therapy for tongue thrust; the seventeen controls had not received therapy. All subjects were randomly selected from orthodontists' records and from therapy files at the University of Utah. All had had Class II, Division 1 malocclusions prior to orthodontic treatment. Three of the 17 therapy subjects and 12 of the nontherapy subjects were found to be currently tongue thrusting. Mean relapse in overjet since the removal of braces was 0.56 mm for the therapy group and 1.94 mm for those who had not received therapy. The relationship between tongue thrust therapy and the amount of relapse was found to be statistically significant at the .02 level of significance. When the therapy and nontherapy groups were combined, those subjects who were predominantly mouth breathers were found to have greater relapse (mean 3.0 mm) than those who breathed principally through the nose (mean 1.3 mm).

Only one study is found in the literature that investigates the results of lip training. Ingervall and Eliasson (1982) gave lip-stretching exercises to a group of 15 children, and withheld such training from a comparable group of ten children, all with a mean age of nine years. The training required from seven to 17 visits to the clinic, with home practice for at least ten minutes each day. Follow-up measurements taken from 11 to 14 months following the beginning of training found a significant difference in lip-lengthening favoring the experimental group.

Summary

Retraining orofacial muscles requires a carefully structured, yet individualized, program applied with skill by a trained clinician to a patient willing to carry out assignments diligently. In order to fulfill the purpose of therapy, the cooperation of several concerned persons is essential. The therapist should be alert to new clinical and research findings, and should modify therapy accordingly.

Therapy can be adjusted to the maturity and needs of children as young as four years of age, or to those of adults of any age. Special attention should be given to opportunities to correct harmful patterns before the mixed dentition period.

The organization of therapy is critical. Early and continuous attention should be given to tongue and lip resting postures, since light, constant pressures are the best movers of teeth, and since those postures affect all other lingual and labial functions. Caution is advised concerning programs that try to achieve normalization of patterns in too short a period of time, as well as those that require an inordinate number of sessions replete with unnecessary assignments.

REFERENCES

Andrianopoulos, M. V., & Hanson, M. L. (1987). Tongue-thrust and the stability of overjet correction. *The Angle Orthodontist, 57*(2), 121–135.

Barrett, R. H., & von Dedenroth, T. E. A. (1967). Problems of deglutition. *American Journal of Clinical Hypnosis, 9,* 161.

Case, J. L. (1975). Palatography and myo-functional therapy. *International Journal of Oral Myology, 1,* 65–71.

Christensen, M., & Hanson, M. (1981). An investigation of the efficacy of oral myofunctional therapy as a precursor to articulation therapy for pre-first grade children. *Journal of Speech and Hearing Disorders, 46,* 160–167.

Christofferson, S. (1970). *The permanency of deglutition changes.* Unpublished thesis, University of Utah.

Cooper, J. S. (1977). A comparison of myofunctional therapy and crib appliance effects with a maturational guidance control group. *American Journal of Orthodontics, 72,* 333–334.

Harden, J., & Rydell, C. M. (1983). Effectiveness of therapy: A study of changes in swallowing habit resulting from tongue thrust therapy recommended by R. H. Barrett. *International Journal of Orofacial Myology, 9,* 5–11.

Ingervall, B., & Eliasson, G. B. (1982). Effect of lip training in children with short upper lip. *International Journal of Orofacial Myology, 10*(3), 17–26.

Ohno, T., Yogosawa, F., & Nakamura, K. (1981). An approach to openbite cases with tongue thrusting habits—with reference to habit appliances and myofunctional therapy as viewed from an orthodontic standpoint. *International Journal of Orofacial Myology, 7,* 3–10.

Overstake, C. P. (1975). Electromyographic study of normal and deviant swallowing. *International Journal of Oral Myology, 1,* 29–60.

Robson, J. E. (1963). *Analytic survey of the deviate swallow therapy program in Tucson, Arizona.* Unpublished thesis, University of San Francisco.

Stansell, B. (1969). *Effects of deglutition training and speech training.* Unpublished doctoral dissertation, University of Southern California.

Subtelny, J. D. (1970). Malocclusions, orthodontic corrections and orofacial muscle adaptation. *Angle Orthodontics, 40,* 170.

Toronto, A. S. (1975). Long-term effectiveness of oral myotherapy. *International Journal of Oral Myology, 1,* 4.

Young, L. D., & Vogel, V. (1983). The use of cueing and positive practice in the treatment of tongue thrust swallowing. *Journal of Behavioral Therapy and Experimental Psychiatry, 14*(1), 73–77.

CHAPTER ELEVEN

RECOMMENDED TREATMENT PROCEDURES

Procedures and approaches described in the preceding chapter are subscribed to by competent professionals. We encourage an open mind toward consideration of their use in the reader's practice. This chapter presents an approach that we have found successful. We modify our own procedures regularly, as we learn of new techniques, and as information from research dictates.

Preliminary Considerations

Chapter Eight treated evaluation, a process that begins on the first visit and continues throughout therapy. That chapter did not discuss the communication that occurs during the first visit of the patient to the therapist's office, communication between therapist and patient, or therapist and patient's parent, in both directions.

Professionalism. The patient should be afforded the respect due by being admitted into a professional-appearing office and being greeted in a warm, professional manner. This does not mean ostentation nor excessive dignity, but proper, intelligent conduct by the therapist, reflecting genuine interest in the patient's well-being. The therapist should acquire the ability to communicate in accurate but lay terminology; patients may be impressed by polysyllables, but seldom respond well to them.

Presence of Observers. In our practices, with few exceptions, patients are required to be accompanied to the therapy room by an observer, be it a parent, spouse, or some other near relative. With children, at least one parent must be present at each session, unless there is an unresolvable conflict that might interfere with rapport or motivation. Even adults are encouraged to bring someone along to the session.

Most specialists who refer their patients to us have good reasons for keeping parents out of the consultation room. It is true that some parents are excessively intrusive or helpful, and that some children are more concerned about the parent's reactions than about those of the therapist, but in most cases the benefits of having them present far outweigh the disadvantages. First, it is easier for the observer to see inside the patient's mouth during home practices than for the patient to do so, even with a mirror. Were the observer not present during the therapy session, s/he would have little concept of what to look for or how to evaluate the practice session at home. Second, as therapy progresses, the patient needs feedback, away from the therapy setting, concerning how well learned patterns are being applied in daily life.

Third, the presence of the parent helps the child to better understand the importance of the therapy. When the parent, as a result of having attended the therapy session, watches the child during home practice sessions and supplies feedback regarding the practice, the child is impressed with the parent's involvement in the whole therapy process. Attitude is all-important in this project, and the improved motivation when parents are informed, interested, and supportive is the factor that usually tips the balance towards habituation of new muscle patterns.

Treating Siblings. Unless siblings are separated by too many years, or do not get along well with one another, two or even three are given therapy together. The advantages of this practice are considerable: half the trips to the therapist, half the time spent observing practice sessions by the parent, and a much-reduced fee from those that would be paid for by separate sessions. If one child progresses more rapidly than the other, the child can mark time for a week or so while the other catches up.

Related to this decision of whether to treat siblings together or separately is the question of whether to see a parent and child simultaneously for therapy. This is generally inadvisable, although there are certainly exceptions. We sometimes encourage parents to do the practicing at home along with the child, but we direct our attention to the child alone during therapy sessions.

The Initial Session. Patient and parents come to the therapist's office with varying degrees of insight into the need for and nature of therapy. You may invite them to ask those questions themselves or you may decide to give them some basic information before encouraging their questions. Before the initial session is over, four objectives should be reached: (1) Motivate the patient to attend to tongue and lip resting postures and to carry out practice assignments faithfully. (2) Provide an overview of therapy, estimating number of sessions and explaining how treatment begins with exercises, then progresses to applications of the exercises in eating, drinking, and saliva handling, includes speech retraining if appropriate, and then habit-strengthening and maintenance. (3) Explain responsibilities of therapist, patient, and parents in therapy. (4) Secure a verbal—and written, if you wish—commitment to therapy from the patient.

1. **Motivation is the Foundation of Therapy.** Referring specialists provide varying amounts of information to the patient they send to the therapist. Children who are merely told by a dentist that they have a tongue thrust have little understanding of the implications. Knowing little, they conjure up strange images of the nature of their problem and of its treatment. They may experience guilt feelings and fear of what therapy may involve. Children old enough to comprehend are best motivated by an understanding of the importance of the correction of their tongue thrust to their future appearance and to the health of their teeth. Younger children must be motivated by some more immediate, tangible reward.

2. **An Overview of Therapy.** It would be unfair and unwise to pop patients into a therapy room and begin assigning exercises without providing them a scope of what lies ahead. The concept of step-by-step progress is emphasized. A warning is voiced about expecting perfection of execution on the first try. The parent is restrained from demanding an immediate change in the patient's swallowing behavior.

3. **Responsibilities of Participants.** The agreement to go ahead with treatment is a pact among at least three parties. In order for success to be achieved, all three must carry out their duties completely. The patient must agree to practice exercises three times daily, and, later in therapy, to carry out all assignments consistently. The parent agrees to observe practices at home as instructed, and to be alert to signs of relapse as the patient eats, drinks, talks, and carries out other daily activities. The therapist agrees to adapt therapy to the needs of the patient, to be clear in teaching and in making assignments, to communicate with the referral source, and to take whatever time is needed to bring therapy to a satisfactory conclusion.

4. **The Verbal/Written Commitment.** After the foregoing is explained to the patient, questions and comments are invited. If any indications of reluctance appear, which is rare, they are probed a bit. If further attempts to motivate are not apparently fruitful, the therapist may ask the patient is s/he would like to delay the beginning of therapy until a later date. A definite commitment should be secured before further sessions are scheduled.

Lip and Tongue Resting Postures

Establishing correct lips-closed resting postures necessitates habitual nose-breathing. At the first session, difficulties the patient experiences in nose-breathing should be ascertained and assessed. If there is any question about the patency of the nasal airway, a referral should be made for an otolaryngologic evaluation. The lips should be capable of resting comfortably against one another, or at least of doing so with reasonably minimal effort. Similarly, the tongue tip should reach the upper alveolar ridge effortlessly. The patient is assigned to be very aware of those resting postures, and is customarily given reminder signs, sticker signals, and sometimes a response counter with which to keep score of efforts to correct postures. A "plus-minus" system is used with patients who are old enough, with a plus earned by keeping tongue on the "spot" and lips closed for a specified period of time, and a minus received when patient or parents become aware of improper resting postures of lips or tongue. Younger patients are given an appropriate reward by parents, daily or weekly, for earning a set number of pluses and not exceeding a set number of minuses.

Attention to lip and tongue resting postures is the only assignment we give the patient on the initial visit. This session is devoted to explanations, motivational procedures, and assignments to initiate awareness of lip and tongue postures.

Outline of Therapy

We view our program as a rational psychophysiological approach aimed at establishing as routine those patterns of muscle movement employed in normal, everyday oral activities. Since these activities are complex acts, they must be broken up into their component parts for therapeutic purposes. The patient is not expected to handle food, saliva, and liquids instantly. The early stages of therapy are directed to exercises that lead to the integrated patterns required for swallowing. Probably the two most important contributions that the program herein has to offer are (1) the concept of disassembling the "motor," taking it apart, machining down some pieces and replacing others, then reassembling it into a smoother-running apparatus, and (2) procedures for achieving changes at a subconscious level relatively quickly.

The various portions that comprise the assignments are the elements of the normal preparatory and swallowing acts, and are referred to as "steps" in therapy; in no sense is this term intended to be interchangeable with "weeks." The length of time required for a given patient to master a step varies with the ability, attitude, and effort of the patient. Ordinarily, one week of practice on each step suffices if the parent supervises, and if practice sessions are carried out three times each day. Each new pattern must be perfected, not simply to the point where the patient is able to execute it, but to the degree that it is performed effortlessly and with little thought required.

There are four general *phases* of therapy, each of which consists of at least two steps.

Phase One: Developing New Muscle Patterns. A helpful over-generalization is that people with tongue thrust over-use the genioglossus muscle, which is the primary tongue protruder, and often over-use the lip muscles, and under-use most of the other muscles of the tongue, as well as muscles active in molar occlusion. In the first phase of therapy, the clinician helps the patient to right these misfunctions; exercises work to deactivate the tongue-protrusion and lip muscles, and to train muscles of mastication and muscles that lift the front, middle, back, and sides of the tongue to function properly. The emphasis is on tongue *movements,* rather than on strengthening muscles, and on perceptions of those movements. The approach, therefore, is sensorimotor. Muscle training begins in the anterior part of the tongue and proceeds posteriorly. As muscle retraining progresses, attention to tongue and lip resting postures continues.

Patients are first taught to raise the tongue tip to the "spot" against the alveolar ridge. This is the new resting place for the tongue tip and its location as a swallow begins. In order for the palatal arch to more easily accommodate the tongue as it lifts, the patient learns to narrow the tongue as it raises. A usually unused reflex, called the "tongue bowl" reflex, is revived or activated, wherein the sides and front of the tongue are raised to form a concavity on the dorsum of the tongue where later in therapy the food is collected prior to swallowing. The patient's attention is called to the masseter and temporalis muscle in a biting exercise, which also serves to strengthen those muscles when necessary.

The patient learns to squeeze the dorsum of the tongue against the palate, with the tip of the tongue remaining on the spot. Once that is learned, s/he learns, easily, in most cases, to swallow water that is first squirted into the buccal cavities then slurped to the back of the mouth, keeping the tongue on the spot, the molars occluded, and the lips spread wide. Further work is done to enhance the patient's awareness of the lifting of the middle and posterior portions of the tongue.

An important element in this phase is teaching the patient to use lip muscles to move material to be swallowed from the front to the back of the mouth. Saliva collects in the lower front part of the mouth; water swallowed a sip at a time has to be moved posteriorly without a tongue thrust; food is introduced into the front of the mouth and must be moved posteriorly before it is swallowed. These posterior movements have heretofore been accomplished with a tongue thrust; now they are the result of sucking the approximated lips back against the labial surfaces of the anterior teeth.

Phase two: Integrating New Muscle Activities into Functional Patterns. If the exercises have been carried out properly, the second phase is easy for most patients. Equipped with the necessary tongue movements, and having learned to sense when the movements are done correctly, they can usually swallow soft food effortlessly and correctly on the first or second try. Sipping and swallowing liquids is learned following continuous drinking. Saliva moving and swallowing ability follows naturally the mastery of exercises that have substituted water for saliva. Eating solid foods is taught as a series of six steps, to be explained later in the chapter as steps are presented in detail. If necessary, speech therapy is provided at this point to correct dentalized linguoalveolar consonants, including the /s/ and /z/.

Phase three: Making Correct Patterns Automatic. Corrected patterns are strengthened by intensive practice of carefully structured assignments. Work on tongue and lip resting postures and on saliva handling, during sleep, follows work on those two aspects of treatment for daytime habituation. Goals are set according to the abilities of the patient.

Phase four: Retaining Learned Patterns. We see our patients for brief, periodic recheck visits until all orthodontic work is completed, and invite their return at any time in the future. If no orthodontic work is to be done, we see them for at least two years following the termination of formal sessions. Visits are spaced progressively farther apart, eventually at six month intervals. Parents are made responsible for detecting signs of relapse during the retention period. Their instructions are simple: Any time they see the child's tongue against or between the front teeth except during the production of the "th" sounds, they are to signal the child—at rest, at mealtime, or during television watching, homework, or at play.

The Program in Detail

Phase one: Developing new muscle patterns

Step One: Tongue and Lip Resting Postures. The patient is shown where to rest the tongue. The "spot" is against the upper gum ridge. In most patients this point is about ¼ inch posterior to the junction of alveolus and teeth. It is explained that this is slightly posterior to the usual resting place for the tongue in non-tongue thrusting people, but the alveolar ridge is an easy landmark to remember and provides a natural barrier to the tongue's forward movement. We explain that some normal swallowers rest the tongue against the lower gums. It will be easier to habituate proper swallowing patterns if the tongue is already resting within the upper arch. At times we press against the spot with the end of a tongue depressor, then ask the patient to do the same. The parent watches closely, and may be asked to touch the child's "spot" with a tongue depressor also.

The therapist then explains the importance of closed-lips resting postures. The lips must touch one another along their entire length. It is not enough that they be *nearly* together at rest. If the upper lip is short and unable to reach the lower lip comfortably, a lip-stretching exercise is assigned. This consists of holding the lower lip immobile (not everted), against the lower gums, opening the mouth far enough to place a tongue depressor edgewise between upper and lower teeth, then stretching the upper lip, using its own musculature, down as far as possible over the entire surfaces of the upper incisors. This position is held for ten seconds, then the patient relaxes for a few seconds. This procedure is repeated ten times, four or five times each day. The period of time necessary for the continuation of the exercise varies from patient to patient, but usually from one to three months is required. Whenever the patient reports it is easy to rest the lips together, and your observations confirm that affirmation, the exercise may be discontinued.

Work on labial and lingual resting postures may need to be supplemented by work on general upper body postures. If the child slumps while sitting, standing, or walking, an open-mouth posture is encouraged. Unless there is a medical reason for the poor posture, normal operant conditioning procedures, if applied consistently over a long period of time, are effective in achieving better postures.

Ordinarily rest posture work continues for at least four weeks after exercises begin. Attention to them is expected of the patient throughout the entire therapy period.

Lip Strengthening Exercises. We work on lip strengthening very rarely. The lips need only have enough strength to remain together at rest and to suck back against the teeth during the posterior movement of food, liquids, and saliva. Nearly always the exercise that teaches the patient to do the sucking back is effective without any strengthening exercises. Nevertheless, a variety of such exercises is available to the clinician. We list some of the more common ones here:

1. *Resistance Device:* Holding a button between the closed lips and the front teeth, against the child's or the parent's efforts to pull it free with an attached string.

2. *Bilabial Opposition:* The lips are squeezed together tightly in a forceful exaggeration of lip closure.

3. *Button or Washer Pullup:* A string is tied to a paper clip. On top of the paper clip are stacked increasing numbers of buttons or washers. The patient holds the free end of the string to the gum ridge with the tongue, bends forward and, reaching down with open lips, grasps and lifts string with lips, repeating until a two-foot length of string is collected into the mouth.

Exercises to Reduce Tonus. Most patients referred to orofacial myologists have an overjet. If the overjet is severe, the lower lip often rests against, and in function seals against, the upper teeth. A hypertonicity may result, which interferes with the proper lips-together resting postures desired in therapy. Some exercises to combat this unwanted tonicity are:

1. Stir a teaspoon of salt into a glass containing four or five ounces of hot water. The patient sips enough of the solution to fill the vestibule between the lower lip and lower teeth. Hold for ten seconds, spit out, and repeat until the solution is gone.

2. Do the above exercise, but with air instead of salt water.

3. Place a dampened cotton dental roll into the vestibule. Hold in place a minimum of an hour each day; several hours are better.

No other exercises are assigned during the first session. The patient is provided an instruction sheet with a practice chart and urged to read the instructions carefully before carrying them out. If there is a question about how to do an exercise, if the instruction sheet is lost, if practices are missed, or if any discomfort results from having done an exercise, the patient is to phone the therapist.

NEXT APPOINTMENT _____

STEP ONE Date: _____

INSTRUCTION TO THE PATIENT

1. Resting positions of the tongue and lips. Try to be very aware of your tongue and lips. Keep the tongue resting against your upper gum ridge, and keep the lips closed as much as possible. Your upper and lower teeth should not be together at rest. Use a plus/minus system. Whenever you keep your tongue and lips in their proper positions for about ten minutes, record a "+" on either a pad of paper you carry with you, or on a counter. Whenever you find your tongue or lips not in their proper resting postures, record a "−." Use the chart below to summarize the number of pluses and minuses at the end of the day.

2. Gum chewing. Best: Don't chew any at all. Next best: Chew sugarless gum, but always with your lips closed, and being careful never to push the gum, or your tongue, against your front teeth. Next best: Chew regular gum, in the same way—no pushing against the front teeth. Worst: Chew bubble gum. Bubble gum chewing encourages tongue thrusting. Don't chew bubble gum!

3. Use the reminders I have given you, and any others you may think of, to remind you to think about your tongue and lips. In addition, choose at least two people to be your "signal-givers."

4. If you were assigned to do "lip stretching" today, do it for ten seconds, ten times each practice, three practices each day. Keep a record on the chart below.

RESTING POSTURE
Plus-Minus

Day:	1st	2nd	3rd	4th	5th	6th	7th	8th
Number of +'s								
Number of −'s								
+ or − day								

Day:	9th	10th	11th	12th	13th	14th	15th	16th
Number of +'s								
Number of −"s								
+ or − day								

LIP STRETCHING

Day:	1st	2nd	3rd	4th	5th	6th	7th	8th

Day:	9th	10th	11th	12th	13th	14th	15th	16th

Rest Posture Reminders. Most patients need something very visible or tangible to enhance their awareness of tongue and lip postures. From a list of reminders compiled at a rap session of the International Association of Orofacial Myology several years ago, here are a few:

1. Post little paper signs around the house, at eye level.
2. Use stickers and decals.
3. The patient writes "TOS" (tongue on spot) or "LC" (lips closed) on hand.
4. Wear a ring on a finger not usually used for a ring.
5. Wear a tiny orthodontic elastic on the tongue tip during certain portions of the day.
6. Place reminder dots on utensils, pencils, and notebooks, and around the house.
7. Wear a watch upside down, or on the other wrist.
8. Carry a small note pad on which to keep score of times with proper postures and times with improper ones.
9. Carry something unusual in your pocket.

On the instruction sheet is a warning about gum chewing. Rapid gum chewing, particularly bubble gum, can foster more tongue thrusts per minute than any other stimulus. It is best if the patient does not chew gum. If this is too severe a restriction, instruct the patient to chew it carefully, lips always closed, and tongue never touching the front teeth during chewing.

Step Two. The patient has had one or two weeks to begin to habituate tongue and lip resting postures and to work on lip exercises if they were deemed necessary. Check to see if the tongue is resting in the correct place. Ask whether there have been any difficulties, and have the patient demonstrate exercises that were assigned. Look at the practice chart to see whether it has been kept, and, if so, how consistent the practicing has been. Four exercises are assigned at this point. Two of them involve placing the tip of the tongue on the spot during the exercise, but each of the two accomplishes something additional. A third exercise focuses on molar occlusion, and the fourth on the tongue bowl reflex. Note the instructions given the patient on the instruction sheet.

1. Tongue Tapping. The purpose of this exercise is to stimulate the involuntary reflex, similar to a grasp reflex, that depresses the middle portion of the tongue in response to a stimulus. Have the patient say "ah" to get the tongue in proper position. The tongue tip is to rest against the lower gums. Tap the middle of the tongue with a tongue depressor with medium pressure. About ¼ inch of the tongue depressor should contact the tongue. Slant the tongue depressor downwards. Encourage the patient to relax the tongue. Continue the tapping long enough to demonstrate proper procedure, then have the patient do so. This is to be continued during each of the three practices each day for one minute. Warn the patient not to hold the tongue depressor too low; it must tap the tongue at an angle. The stick must leave the tongue between tappings. Often the depression in the tongue occurs during these first few minutes in your office. At other times it takes several days to elicit the response properly. In almost all cases the bowl reflex is clearly evident by the time the patient returns for the next visit. Practice for one minute. This exercise is usually continued for two weeks, even though the next session is scheduled for the succeeding week.

2. Tongue Narrowing. If the tongue is to fit easily into the palatal arch during swallowing, it must narrow slightly as it lifts. This exercise gets both the lifting and the narrowing started. Demonstrate the exercise. Protrude your tongue, keeping it as narrow and pointed as possible. Keep it up off the lower teeth and lower lip as you do so. The front of the tongue should not be touching any structures. Very slowly retrude the tongue, lifting its tip to the "spot." Bite down slowly, keeping the tip of the tongue on the spot. Then open slowly and again protrude the tongue. Have the patient try it, watching a mirror closely. At first it is often difficult to keep the tongue from resting on the lower lip or teeth. Encourage the patient to do as well as possible. The exercise is most effective if done slowly. Continue the narrowing exercise for two minutes each practice.

3. Biting. Before teaching the biting exercise, be sure of two things: (1) The exercise is needed. Some people with tongue thrust already do an adequate job of occluding the molars during swallowing. (2) Inter-relationships between corresponding molars on both sides of the arch are such that work on biting will not do any harm, and that bilateral symmetry of masseter muscle contractions is a reasonable expectation.

In some patients, the muscles of mastication are underdeveloped and need strengthening. In most patients, however, the muscles are fine, they just haven't been using them to occlude the molars for swallowing. A tongue-teeth seal has been doing the work the biting muscles should have been doing. The purpose for those patients is to enhance awareness and to begin to build a pattern of biting for swallowing, rather than to strengthen muscles.

NEXT APPOINTMENT _____

STEP TWO   Date: _____

WATCH THE MIRROR CLOSELY FOR ALL EXERCISES!!

Practice the following exercises three times each day, seven days each week. If you miss any two practices during the week, call me no later than the day after you miss the second time. Please don't wait until the day of your appointment to phone. If I am not there, please leave a message with the secretary.

1. **TONGUE TAPPING.** Let your tongue rest on the floor of your mouth, with the tip touching your lower gums. Tap the middle of the upper part of your tongue for *one minute.* Slant the stick down toward your tongue, allowing only the very end of the stick to touch the tongue. Keep your tongue relaxed.

2. **TONGUE NARROWING.** Stick your tongue out, keeping the tip up. Make it as narrow as you can. Slowly draw it back into your mouth, until it is back far enough to touch the "spot." Keep it narrow as you move it back. Slowly bite down, keeping tongue narrow. Do this for *two minutes* each practice.

3. **BITING.** Place your fingers on each side of your face, just in front of your ears. Bite and feel the muscles tighten. Keep them tight for ten seconds, then relax. Try to keep the tip of your tongue on the "spot" while you do this exercise. Your parent can place a hand on one of your cheeks while you touch your other cheek. Do this *10 times* each practice.

4. **OPEN and CLOSE.** Suck your tongue up into the roof of your mouth, as if you were going to pop your tongue. Open your mouth wide and hold with your tongue sucked up for *5 seconds.* Then bite down slowly on your back teeth, keeping your tongue sucked up as you bite down. Hold for *five more seconds.* Do this *10 times each practice.*

5. **LIP STRETCHING.** Continue this if you were assigned to do it. Ten seconds, *10 times each practice.*

6. **PLUS–MINUS.** for tongue and lip resting postures. 15 minutes = one plus, Minuses are instant. Try to get at least ____ +'s and no more than ____ –'s.

DAY	1st	2nd	3rd	4th	5th	6th	7th
1. TAPPING							
2. NARROWING							
3. BITING							
4. OPEN–CLOSE							
5. LIP STRETCHING							
6. NO. OF +'S							
NO. OF –'S							
+ OR – DAY?							

Place patient's hands on your cheeks and bite firmly. Then have the patient place his/her hands on their own cheeks, bite firmly, evaluate symmetry of tightness of muscles, hold for ten seconds, then relax. The patient repeats this 10 times each practice, or for two minutes.

4. *Open and Close.* The purpose of the "open and close" exercise is to get the tongue, with its tip in the proper place, to squeeze up into the palatal arch. It is a variation on the old "tongue popping" exercise, and is best taught using tongue popping as a starter. The patient pops the tongue as loudly as possible, watching the mirror to see that just before the tongue is forcefully pried loose from the roof of the mouth, the frenum under the tongue is seen to stretch, and the under part of the tongue to suck up against the palate. Next, a tongue pop is initiated, but instead of pulling the tongue down, the patient holds it in its sucked-up position, noticing once more how the lingual frenum stretches. The teeth are $1/2$ to $3/4$ inches apart. Keep the tongue in that position for five seconds, then bite down slowly, keeping the tongue sucked up as you do so. When the teeth are together, keep the tongue up for another five seconds. Repeat this ten times each practice, or for two minutes, whichever is less.

If the patient is unable to learn this exercise, assign very slow tongue pops for two minutes each practice.

5. *Lip Exercises* continue if they were assigned last session.

6. *Systematic Work on Establishing Correct Tongue-Lip Resting Postures* continues. Set a minimum number of pluses to achieve each day, and a maximum number of minuses, based on the patient's record for the previous week.

Step Three. Almost always the tapping and narrowing exercises are continued for an additional week. Check these and other exercises assigned during the previous session to see whether they have been carried out as instructed. If not, reassign them for an additional week, after having taught again their correct execution.

The third exercise assigned at the third step (the first new one) is "Slurp and swallow." This teaches a correct swallow in its simplest form. Water is squirted into the patient's mouth, slurped back along the sides of the tongue, and swallowed while the tongue remains within the upper arch. Slurping is just an easy way to get the liquid to the back of the mouth for swallowing. The patient watches the mirror, keeps the back teeth together during slurping and swallowing, and the lips wide apart. Lips are held apart for two important reasons: (1) To allow the patient to see that the tongue does not come forward, and (2) to disassociate the lips from the swallowing process. In most tongue thrusting, the forward tongue, the separated upper and lower teeth, and the tightened lips together form a seal that allows the food or saliva to be moved posteriorly and swallowed. Keeping the lips wide apart for several weeks of exercises and assignments helps break up this triad of actions.

Explain to the patient that the water is slurped back along the sides of the tongue, not along its dorsum. The harder the slurping action, and the quicker the swallow that follows it, the easier is the exercise.

During ten of the twenty repetitions of this exercise assigned the patient, instead of watching a mirror, the patient places a plastic straw of small diameter across the mouth, usually just behind the cuspids. At times it is better to place the straw mesial to the cuspids, particularly when the arch is narrow. The objective is to accomplish the slurping and swallowing without the tongue's touching the straw. In order to do this, the patient must keep the under side of the tongue from lowering or moving anteriorly. At times, keeping the tip of the tongue on the spot results in forward movement of the under portion of the tongue. This variation of the exercise helps prevent that. Assign five tries with the mirror, then five with the straw, then five again with the mirror and five more with the straw. Explain the purpose of each portion of the exercise.

The fourth exercise assigned during step three is a whistling or a strident /s/. Its purpose is to give the patient time to attend to the sensation of making contact between the sides of the tongue and the lateral alveolar processes. Each corrected swallow should begin with the tongue tip on the spot. As the rest of the tongue squeezed up, the lateral margins of the tongue should not press against any teeth within the upper arch except for the molars.

Most patients are unable to whistle without puckering the lips, but nearly all learn within a couple of minutes to produce some semblance of a whistle, by first producing an /s/, then slowly drawing the tongue up tighter within the arch, and curling the tip back slightly. Even if a whistle does not result, the tongue is in its proper position if the patient knows its sides are not contacting the cuspids or bicuspids. Since most of your patients have malocclusions that permit viewing of the tongue with teeth occluded, you can tell whether the sides of the tongue are in the proper position by simply looking. If a reasonably sharp /s/ sound, or a whistle, are being produced, and the tongue is not visible, its sides are probably placed correctly.

Have the patient hold this position, whistling or saying the /s/ from

NEXT APPOINTMENT _____
STEP THREE   Date: _____

WATCH THE MIRROR CLOSELY FOR ALL EXERCISES!

1. and 2. **TAPPING** and **NARROWING**. Continue both exercises for *one minute each.*

3. **SLURP** and **SWALLOW**. Put your tongue tip on the spot and gently squeeze the top part of the tongue up against the roof of your mouth. Bite down hard and spread your lips wide. Squirt water into the sides of your mouth. Slurp back hard, and swallow. Do this *five times.* Then place a straw behind your upper cuspids. Bite down, spread your lips wide, keep your tongue on the spot. Squirt water, slurp hard, swallow. Do this *five times.* Don't Let The Tongue Touch The Straw! Then do *five more* slurps without the straw, then *five more* with it. Do a total of *twenty slurps each practice.*

4. **WHISTLING**. Purpose: To Feel the *Sides* of Your Tongue Against your Upper Gums. Say "ssssssss." As you hold the "ssssssss," curl the tip of your tongue slowly back and draw its sides up tighter against your gums. Try to make a whistling sound. When you have it as sharp as you can get it, keep it there for at least *two minutes.* Remember, this is a FEELING exercise.

5. **LIP STRETCHING**. Continue this exercise if you have been assigned to do it.

6. **RESTING POSTURES**. Try for at least _____ +'s and not more than _____ -'s. 15 minutes = 1+.

DAY	1st	2nd	3rd	4th	5th	6th	7th
1. and 2. TAPPING AND NARROWING							
3. SLURP AND SWALLOW							
4. WHISTLING							
5. LIP STRETCHING							
6. NO. of +'s							
NO. of -'s							
+ or - DAY?							

time to time, for two minutes. This is an exercise many children enjoy doing during the day. Whistling tunes is legal, just so the child's attention is on the *feeling* of having the sides of the tongue against the upper gums.

Lip exercises and work on tongue and lip resting postures continues.

Step Four. Previously assigned exercises are checked. In most cases they are all discontinued at this session. The first new exercise taught is "trapping." In order for a non-tongue thrust swallow to be effective, the seal between sides of tongue and alveolar ridge must be tight as the dorsum of the tongue is raised against the palate. Leaks in the system foster a return to the very efficient, but harmful, tongue thrust pattern.

1. Trapping. By now the tongue readily responds to stimuli by assuming the spoon, or bowl shape. Demonstrate the exercise by protruding your tongue slightly and raising its tip and sides. Squirt water into the pocket on top of the tongue, then slowly draw it posteriorly and to the spot, without letting the tip or sides drop as you do so. Bite down, spread your lips wide, and swallow, keeping the tip on the spot. Then have the patient try it. If the concavity is not obvious, squirt anyway; it will usually appear. The patient watches the mirror to make sure the sides and tip of the tongue remain elevated, and that no water leaks out the sides or over the tip as the tongue moves back. When this is accomplished, explain that from time to time you want the trapping tested by having the patient lean over, lips spread, teeth together, water trapped on the tongue, without losing any water out of the mouth. Then lift the head and swallow.

2. Back-Middle-Tip. Lifting of the middle and back portions of the tongue is the focus of this exercise. Rather than having the patient squeeze the tongue up sequentially from front to back, as occurs during swallowing, the order is reversed, for two reasons: (1) Raising the back of the tongue first, then the middle, then the tip, allows the procedure to be watched in a mirror. Were the tip raised first, then the middle, the front part of the tongue would obscure the remaining parts. (2) Lifting the back of the tongue first requires more effort than lifting it when the front part of the tongue is already raised. More benefit is realized by muscles that raise the posterior tongue when the front portion remains low in the mouth until after the back and middle parts are raised.

The first step in teaching the patient is to have him/her say "ah" while watching the mirror. Call attention to the low posturing of the entire tongue, and to the position of its tip behind the lower gums. This is the position the tongue should be in when the patient squirts water back into the throat area to begin the exercise. Then the /k/ is repeated, slowly, a few times while the patient notes the raising of the back of the tongue. Next, the back of the tongue is raised as if to produce the /k/, but it remains against the velum instead of dropping, as the middle part of the tongue is raised, then the tip. Explain that the tongue is being rolled up against the palate from back to front.

Next, have the patient tip the head back, place the tongue in the position for saying "ah," then squirt some water well back into the throat. The patient thinks of the /k/ sound, lifts the back of the tongue, then the middle, then the tip, in clearly defined stages, bites down, spreads lips wide, and swallows, keeping the tip of the tongue on the spot. This is repeated 20 times during each home practice.

Patients who are unable to learn this exercise can substitute the simple /k/ exercise, producing a /k/ several times in a row, each time keeping the back of the tongue elevated for a longer period of time. After a number of /k/ productions, some water can be squirted into the posterior mouth and swallowed with back of tongue raised and teeth still apart. Do this for two minutes.

3. Squeaky and Quiet Sucking. This is the last formal exercise to be taught in the program. Its purpose is to teach the very important, and often neglected, skill of moving saliva to the back of the mouth by sucking the lips back against the teeth, with tongue on spot. People with tongue thrust usually accomplish the posterior movement of saliva and food by creating an anterior tongue-teeth seal. This seal must be replaced by a lips-against-teeth action. It is most easily taught by having the patient make a squeaking sound while sucking water back. It is the same sucking action used when taking liquid through a straw, except that the lips are so close together the squeaking results. Alternating a very slight separation of the lips with closure while sucking back helps the patient learn the technique. Distinguish between the squeaking sound required and the slurping done in the earlier exercise. The lips are closer together, and are sucking back against the teeth.

With your teeth together and lips spread, squirt some water into the anterior vestibule and demonstrate. When the water is sucked back, spread the lips wide and swallow, keeping the teeth together and the tongue tip on the spot. Then have the patient try. This is difficult for some children and may take a few minutes to teach. See if the parent is able to do the exercise; if not, teach the parent, who will then be able to help the child at home.

NEXT APPOINTMENT _____

STEP FOUR   Date: _____

WATCH THE MIRROR CLOSELY FOR ALL EXERCISES!

1. **TRAPPING.** (1) Stick your tongue out. (2) Raise the tip and sides of your tongue. (3) Squirt water into the "pocket" your tongue has made. (4) Keeping the sides of your tongue up, slowly lift your tongue back and up into the roof of your mouth. As you do this, bite down on your back teeth. (5) Spread your lips wide, keeping the back teeth together. (6) Tilt your head forward to test your "trap" for leaks. (7) Head straight. (8) Squeeze the middle of your tongue up, and swallow. Watch the mirror closely! Do *20 each practice.*

2. **BACK–MIDDLE–TIP.** Keep your head tipped back as you do this exercise.
   (1) With your tongue resting on the floor of your mouth, squirt water back into your throat.
   (2) Keeping the tip of your tongue down, behind your lower gums, raise the back of your tongue as though you were going to say a "K" sound.
   (3) Without dropping the back of the tongue, squeeze the middle part of the tongue up, then the tip. Be sure the tip ends up on the "spot".
   (4) Bite down on your back teeth, keep your lips wide apart, and swallow. Do not let your tongue show against your front teeth. *20 times each practice.*

3. **SQUEAKY** and **QUIET SUCKING.** (beginning saliva moving) (1) Put your tongue on the spot, and bite down. (2) Spread your lips wide and squirt water into your mouth. (3) Close your lips gently and suck back, making a *squeaking,* not a slurping, sound. (4) Spread your lips wide again, and swallow. Make sure your tongue tip stays on the spot during the whole exercise. Do this *10 times.* Then repeat, sucking the water back quietly, without a squeak. Do this *10 times.*

4. **RESTING POSTURES.** Continue your +, – work. Try to get at least ____ +'s and no more than ____ –'s each day.

DAY	1st	2nd	3rd	4th	5th	6th	7th
1. TRAPPING							
2. BACK–MIDDLE–TIP							
3. SQUEAKY and QUIET SUCKING							
4. NUMBER OF +'S							
NUMBER OF –'S							
+ OR – DAY							

The quiet part of the exercise is easier, but does not permit the therapist to tell whether the sucking action is being carried out correctly. It is done in the same way, except with the lips a little more tightly approximated so that no squeaking sound is produced. Assign ten squeaky sucking tries, followed by ten quiet ones, for each practice. Be sure the patient watches the mirror as the lips spread before the swallow occurs, to determine whether the tongue has moved forward against the teeth during the sucking-back action.

Nearly all patients, except the very young ones, are able to learn the squeaky and quiet sucking exercises quite quickly. A few have trouble with the squeaky part. If they cannot learn it, just teach the quiet sucking.

Lip exercises and resting posture work continues, in most cases, for at least another week.

Phase Two: Integrating Movements into Patterns

Step Five: For most patients, exercises for training muscles to function properly for correct handling of food, saliva, and liquids are now completed. If the patient is able to demonstrate proficiency in all exercises assigned to date, it is time to integrate the muscle movements into patterns for everyday use. This step teaches swallowing of soft food, continuous drinking, and saliva swallowing.

1. Beginning Food Swallows. You may wish to eliminate this step and teach chewing and swallowing directly. We have found soft food a helpful introduction to eating, because it is simple and provides a pleasant success experience for a week. Using a tongue depressor or spoon, place a small amount of pudding or yogurt in the middle of the tongue. As the patient watches the mirror, s/he lifts the tongue tip to the spot, watches the sides of the tongue to be sure they touch the upper gums rather than the upper teeth, bites down, spreads lips wide, squeezes the tongue up, and swallows. Most patients do this correctly the first try. Assign 20 swallows or more each practice, three times a day. Any food that does not need to be chewed can be used for home practice.

2. Continuous Drinking. We teach patients to drink so carefully that they are doing so abnormally, and we explain this to them and give them the reasons for doing so. Normally, (1) we drink with front of tongue free, not touching teeth nor tissue. (2) Our teeth are held apart and (3) we tip the head forward slightly to begin the drink. While the patient is learning to drink, these three procedures work against eliminating the tongue thrust. The tongue needs a reference point; swallowing is easier if the jaws are kept stable through molar occlusion; tipping the head forward makes the water travel uphill to reach the throat. So the patient is instructed to hold the head straight, keep the tongue tip on the spot, and keep the teeth together while learning to drink. We have patients continue drinking in this manner for several weeks, then allow the teeth to part. Have the patient take three or four swallows in a row, take the glass away from the mouth, take four more swallows, and proceed in this manner until a seven or eight ounce glass of liquid is finished. Do this at least three times a day, or for every drink during the day.

3. Moving and Swallowing Saliva. The squeaky and quiet swallowing exercise has taught the patient how to handle saliva. To stimulate saliva production, have the patient keep a sugarless mint in the cheek, three times a day, until the mint dissolves. Whenever enough saliva is produced to deal with, keeping the tongue on the spot and the lips closed, bite down, suck the lips back against the teeth, squeeze the tongue up against the palate, and swallow. Emphasize the importance of keeping the tongue back during the posterior movement of saliva. Most patients get from 75 to 100 good saliva swallows from each mint. After the mint is gone, the taste it leaves can serve as a reminder to continue to handle saliva in this manner throughout the day.

This sugarless mint use is continued for several weeks, and in some cases for several months. The patients enjoy it, and it helps keep the mind on saliva swallowing. By now, lip exercises and scorekeeping for lip and tongue resting posture work may be terminated, in most cases. Some incompetent lips may need as much as another month of work, however.

Step Six. Two new functions are taught at this point: handling food that must be chewed, and drinking a sip at a time. Work on saliva swallows continues, and speech work is begun if necessary.

1. Chewing and Swallowing Food. Explain to the patient that there is one important principle to be remembered while eating: At no time during eating should the tongue ever touch the front teeth, for any reason. Broken down into steps:

a. As food comes toward the mouth, keep the tongue back. It need not be on the spot, just so it is behind the front teeth. Protruding the tongue to meet the food sets it up to thrust during chewing and swallowing.

b. Take a reasonable-sized bite. Too small a bite is difficult to sense; too large a bite ends in a bolus too large to fit within the pocket on the dorsum of the tongue.

c. Chew with lips closed, moving the tongue laterally, or in a circular

NEXT APPOINTMENT _____

STEP FIVE   Date: _____

WATCH THE MIRROR CLOSELY FOR ALL EXERCISES

1. **BEGINNING FOOD SWALLOWS.** Place a small amount of soft food (pudding, yogurt, jello, apple sauce, ice cream, etc.) in the middle of the top of your tongue. Lift your tongue up, and swallow. See if you have swallowed all the food. WATCH THE MIRROR CLOSELY. Nothing should come forward as you swallow. *20 times.*

2. **CONTINUOUS DRINKING.** Right when your hand first touches the glass, do three things:
   (1) Put your tongue tip on the spot.
   (2) Bite down on your back teeth.
   (3) Be sure that your head is straight.
   Stay like that all through your drink. Take 4–5 swallows, continuously, then take the glass away. Repeat until you have drunk and average-sized (8 oz.) glass of liquid. *Do this at each of the three practices.*

3. **SALIVA MOVING AND SWALLOWING.** At least three times a day, hold a sugarless mint inside your cheek to get saliva flowing. Keep your tongue tip on the spot. Whenever you have enough saliva to swallow, bite down on your back teeth, suck back with your lips closed, and swallow. *Do not bite or chew the mint.* Keep repeating this action until the mint is gone. After it is gone, let the taste that remains remind you to keep swallowing your saliva right.

DAY:	1st	2nd	3rd	4th	5th	6th	7th
1. BEGINNING FOOD SWALLOWS							
2. CONTINUOUS DRINKING							
3. SALIVA MOVING AND SWALLOWING							

motion, but without touching the anterior teeth. During chewing, the lips move the food posteriorly, the cheeks move it centrally, and the tongue moves it onto the molars for mastication. If the tongue attempts to take over the functions of lips and cheeks, a tongue thrust results.

d. As chewing ends, carefully place the tip of the tongue on the spot, bite down on the back teeth, suck the lips back against the teeth, then, watching the mirror closely, spread the lips wide, squeeze the tongue up from front to back, and swallow. During swallowing, nothing should come forward; not the tongue, nor saliva, nor food.

e. If food remains in the anterior part of the mouth after swallowing, do not clean it up by sweeping the area with the tongue. Instead, close, suck back again, and repeat the swallow. Or, just take another bite of food and wait for final cleanup until eating is over. Swishing or brushing can take care of any food remaining in difficult places between teeth.

It is helpful to have a sign made, either by the clinician or by the child, listing the above steps:

TONGUE BACK AS YOU TAKE A SMALL BITE
CHEW WITH THE LIPS CLOSED
GATHER FOOD RIGHT
SWALLOW WITH TONGUE ON SPOT
NO "CLEAN-UP" TONGUE THRUST

Assign the child to eat any five foods correctly each day, or to eat everything right at dinner, according to the child's preference. Some may prefer to eat everything correctly right away. The mirror work is essential. If it is embarrassing for the teen-ager to watch a mirror during meals, for a time ask the parents' permission to have him/her eat away from the table. It is dangerous to omit the mirror watching. Even adults must use the mirror.

2. Sip at a Time Drinking. Drinking a sip at a time involves exactly the same procedure as the quiet sucking exercise. The only difference is the water is sipped instead of squirted into the mouth. Place the tongue on the spot, bite down, take a small sip, then suck the water back by sucking the lips against the teeth. Watch the mirror, spread lips wide, squeeze tongue up, and swallow. When the lips are spread, look at the anterior lower vestibule; no water should remain there. Assign 20 sips, three times a day, or sipping at two meals a day. Whatever else the patient drinks during the day should be drunk continuously and correctly. The patient is taught to drink from a fountain and through a straw with head tipped forward, teeth apart, and tongue on the spot.

NEXT APPOINTMENT _____

STEP SIX   Date: _____

HANDLING FOOD THAT MUST BE CHEWED

1. **CHEWING AND SWALLOWING.** (1) As you bring the food toward your mouth, don't let your tongue reach out for it. Keep your tongue low, resting behind your bottom teeth. (2) Take reasonable-sized bites. (3) Chew with your lips closed, moving the food back into your mouth as you chew. Keep your tongue away from your front teeth as you chew. (4) When you have chewed the food long enough that it sticks together, lift your tongue to the "spot," bite down, and suck your lips back against your teeth. (5) Spread your lips wide, squeeze the tongue up, and swallow. Use the back muscles of your tongue to swallow. Eat at least five different foods correctly each day. Write down the five foods you eat right each day on the chart below.

1st DAY: _____ _____ _____ _____ _____

2nd DAY: _____ _____ _____ _____ _____

3rd DAY: _____ _____ _____ _____ _____

4th DAY: _____ _____ _____ _____ _____

5th DAY: _____ _____ _____ _____ _____

6th DAY: _____ _____ _____ _____ _____

7th DAY: _____ _____ _____ _____ _____

2. **"SIP AT A TIME" DRINKING.** (1) Tongue on spot and teeth together. (2) Take a *small* sip of water. (3) Close your lips and suck the water back. (4) With your teeth still together, spread your lips wide and swallow. *15 times each practice.*
3. **CONTINUE USING SUGARLESS MINTS.**
4. **SPEECH.** _____
   _____
   _____
   _____

DAYS	1st	2nd	3rd	4th	5th	6th	7th
2. "Sip at a time" drinking							
3. Mints							
4. Speech							

3. Continue Using Sugarless Mints.

4. Speech Work. By now the patient will have had several weeks of practice with tongue and lip resting postures. The "spot" will be a comfortable place for the tongue, and work on the correction of any dentalized linguoalveolar sounds will be facilitated. If only the /s/ and /z/ are dentalized, provide whatever therapy you would ordinarily use for their correction. If other consonants are dentalized as well, or, in other words, if there is a feature error involving linguoalveolars, approach the problem as such.

For children able to read, have them exaggerate the posteriorization of the linguoalveolars by imagining there is a tiny coil spring connecting the tip of the tongue to the spot. As they read, the spring keeps drawing the tongue tip back up to the spot every chance it gets. The speech will sound a little retroflexed if this is done correctly. During the second week the speech can be normalized. Progress from reading to conversational speech in steps. For children unable to read the procedure is the same, but conversation is used instead.

Informal work on resting postures continues, with new reminders introduced according to the needs of the child.

Phase Three: Strengthening Patterns.

Step seven. Habit strengthening is the focus for the next few sessions. Check all swallows carefully, and be sure parents or other watchers are able to tell the correct from the incorrect. During continuous drinking, the tongue should not be seen while the glass approaches the mouth. While the glass is at the mouth, the lips should be quiet, should not protrude, and the head should remain straight, tipping back slightly as the drink ends.

1. During eating and drinking the parent should not see the tongue at any time. The lips should move during chewing, but not protrude. When the lips spread, the tongue should not be seen darting back into the mouth from against the teeth. After the swallow, there should be no signs of food gathering with the tongue. The patient is assigned to watch the mirror, and spread the lips during swallowing, for all food eaten at home. A chart is provided.

2. Speech work continues as necessary, as does attention to resting postures.

3. Television watching or reading. The patient is now able to eat and drink correctly while thinking about doing so. Eventually all swallows

must be done correctly when attention is not on them. This assignment provides a step toward that goal: part of the patient's attention is on swallowing while part of it is on some other activity. Either a sip of water, or a bite of food is taken and swallowed while the patient makes a conscious effort to continue reading or watching a television program. Make the food or liquid last a half-hour. This is an assignment most patients enjoy, and it is not difficult to have them continue it for several weeks. Explain the importance of achieving automaticity of all swallows.

4. Continue using mints.

5. Nighttime swallows. Discuss swallowing at night. We probably swallow 50 to 100 times during sleep, or perhaps more. If breathing during sleep is principally through the mouth, years of habit strength associating open mouth rest postures with tongue thrusting will foster the perpetuation of the tongue thrust. Even though the number of swallows during sleep is relatively small, eight hours or more of improper resting postures and incorrect saliva swallows will interfere with total habituation of corrected patterns. Conversely, if swallows that occur during the night can be corrected, that same number of hours of corrected subconscious swallowing can contribute greatly to the overall consistency of swallows during the day.

We explain that we are going to apply some principles of self-hypnosis in this assignment. During a "twilight zone" between the waking state and deep sleep the patient narrows conscious thought down to the areas of resting postures of tongue and lips and correct swallowing of saliva. S/he repeats, over and over, a suggestion, "I will swallow right all night," or "I will keep my lips closed, my tongue on the spot, and swallow right all night." As sleepiness increases, the thought continues to be repeated, and the patient goes to sleep with the suggestion well implanted in the mind. Many report that the tongue almost seems glued to the spot when they awaken in the morning. Nearly all patients report success with this procedure. Consistency in its application is important; impress the patient with the importance of doing it each night. Before going to bed, the patient is to do 25 "quiet sucking" exercises, repeating "I will swallow right all night" after each time.

Step Eight: Work on habituation continues. An important chart is utilized, which assists the patient in assessing progress and in determining areas that need added attention. The television or reading assignment, speech work, and sugarless mint use all continue.

1. *Progress chart:* The chart lists skills the patient has acquired along

NEXT APPOINTMENT ——————————
STEP SEVEN   Date: ——————————

HABIT STRENGTHENING

1. **EATING.** Eat *everything* right. For everything you eat at home, watch the mirror and spread your lips wide as you swallow. On the eating chart below, write a "+" if you ate everything right during that meal; a "P" for "part" if you ate part of the meal right; and a "−" if you forgot to eat right during that meal.

2. **SPEECH.**——
——

3. **TV or READING:** Drink about 8oz. of liquid, a sip at a time, while reading or watching TV. Keep your attention on the TV or book. Make the 8oz. last a half-hour.

4. **MINTS.** Continue using sugarless mints.

5. **NIGHT SWALLOWS.** Right before you go to bed at night, do 25 Quiet Sucking Exercises as you learned in Step 4. After each swallow, think to yourself: "I will swallow right all night." Then, as you go to sleep, keep repeating this same phrase in your mind. Be sure that your lips are closed and your tongue is on the spot as you go to sleep.

1. EATING CHART

	First Week			Second Week		
	BREAKFAST	LUNCH	DINNER	BREAKFAST	LUNCH	DINNER
1st Day:	————	————	————	————	————	————
2nd Day:	————	————	————	————	————	————
3rd Day:	————	————	————	————	————	————
4th Day:	————	————	————	————	————	————
5th Day:	————	————	————	————	————	————
6th Day:	————	————	————	————	————	————
7th Day:	————	————	————	————	————	————

FIRST WEEK

DAY:	1st	2nd	3rd	4th	5th	6th	7th
2. SPEECH							
3. TV or READING							
4. MINTS							
5. NIGHT SWALLOWS							

SECOND WEEK

DAY:	1st	2nd	3rd	4th	5th	6th	7th
2. SPEECH							
3. TV or READING							
4. MINTS							
5. NIGHT SWALLOWS							

the side, and days along the top. Each night before going to bed the patient, using either percents or some kind of letter score scale, evaluates performance during the day. Items receiving the poorest rating become special targets for attention and improvement the next day. The goal is to raise scores as much as possible on all tasks. This chart is to be kept for two weeks, until the next appointment.

2. *Continue using mints.*

3. *Do TV or reading assignment.*

4. *Continues speech work.*

5. *Nighttime elastics.* The work on night swallows continues. After alternating the quiet sucking exercise with self-message several times, the patient places a tiny (1/4 inch, thin-walled) orthodontic elastic on the tip of the tongue and places the tongue on the spot. The objective is to keep the tongue in that position all night and to awaken in the morning with the elastic still in place and the tongue still on the spot. While going to sleep, the patient repeats the appropriate words over and over.

Most patients are able to succeed at least once during the first two weeks in keeping the elastic in place. We tell them there is no disgrace nor failure in not being able to do so; no law, natural nor man-made, requires that the tongue stay in one place all night long. For many reasons the elastic may be displaced, or swallowed. The assignment is worthwhile even if the elastic is not there in the morning.

Some children or their parents are reluctant to place the elastic on the tongue tip. Neither of us has ever had a patient have any difficulty, and we have both used the elastics for over 25 years. Nevertheless, life goes on without this assignment, and we prefer to omit it if there is concern or anxiety. There are other ways to test the automaticity of the saliva swallow.

Step Nine. Check the patient's swallows for ease of performance and correctness. One of several variations of a squirt test may be used for this purpose. One variation is to have the patient count backwards, beginning with 100, 99, 98, etc., holding the mouth quite wide open while doing so. At variable intervals, squirt water into the mouth. The patient swallows, then continues counting. Note particularly whether the tongue comes forward as the water is being moved posteriorly. This is detectable just as the lips part before swallowing. The tongue may be seen moving back away from the teeth as the lips part, indicating there has been a tongue thrust. You usually have evidence of poor performance by the time the count reaches 75 or so.

NEXT APPOINTMENT _____

STEP EIGHT   Date: _____

WORK ON HABITUATION CONTINUES

1. **CONTINUE TO EAT AND DRINK EVERYTHING RIGHT,** watching the mirror for all food eaten at home.
2. **CONTINUE TV OR READING ASSIGNMENT.**
3. **USE 3 MINTS A DAY.**
4. **CONTINUE SPEECH WORK.** _____

5. **NIGHTTIME ELASTICS.** Before going to bed, do 10 "Quiet Sucking" exercises. After each time, think "I will swallow right all night." Then, place one of the tiny elastics I have given you on your tongue. Continue to repeat the same words over and over again in your mind as you go to bed. In the morning either check yourself, or have someone check you, to see whether the tongue is on the spot, the elastic is still on the tongue tip, and the lips are closed when you awaken. Mark a "+" if the elastic is still on the spot; write "TOS" if your tongue is on the spot, but the elastic is gone: or a "−" if the elastic is gone and your tongue is not on the spot.

1. PROGRESS CHART

Fill out the following chart each night, using either percents or the following abbreviations: A = All the time; NA = Nearly always; ½ = Half the time; S = Some or less than half; and VL − Very little.

DATES:															
1. TONGUE ON SPOT															
2. SALIVA MOVING															
3. SALIVA SWALLOWING															
4. DRINKING															
5. PROPER CHEWING															
6. PROPER COLLECTING															
7. PROPER SWALLOWING															

DAY	1st	2nd	3rd	4th	5th	6th	7th
2. TV or READING							
3. MINTS							
4. SPEECH							
5. NIGHT SWALLOWS							

NEXT APPOINTMENT _____

STEP NINE   Date: _____

SETTING GOALS

1. **CONTINUE WITH NIGHTTIME SWALLOWS.**

2. **CONTINUE WITH SUGARLESS MINTS.**

3. **CONTINUE WITH TV OR READING ASSIGNMENT.**

4. **SPEECH WORK.** _____

_____

4. **USE THE GOAL CHART.** Try to meet each goal, then never drop below it.

GOAL CHART

DATES	TODAY					GOAL					GOAL			
RESTING														
SALIVA														
DRINKING														
EATING														

DATES	GOAL				GOAL				GOAL
RESTING									
SALIVA									
DRINKING									
EATING									

DAY:	1st	2nd	3rd	4th	5th	6th	7th
1. Nighttime Swallows							
2. Mints							
3. TV Reading Assignment							
4. Speech Work							

Forced Tongue Thrust. Another variation is to tell the patient you are going to try to force a tongue thrust, by squirting, having the patient swallow, squirting again, and gradually increasing the rapidity of the squirts until a point is reached where a tongue thrust occurs. Patients who are doing well can swallow correctly as rapidly as you can follow each swallow with a squirt.

The Sidelong Glance. A third variation: Give the patient a wafer. Tell him/her you want to check just one or two swallows. When two have been done correctly, say that it was fine and have the patient finish the wafer as you busy yourself doing some writing or talking to the parent. Watch the patient out of the corner of your eye to see how the eating proceeds.

The Nighttime Elastic. Work continues on this assignment. If the patient has not succeeded in keeping the elastic in place, try to determine why. Ask whether breathing through the nose has been difficult or easy. If anything has to be done to clear the nasal airway, make a referral to an otolaryngologist. Ask a parent to go into the child's room while s/he is asleep and check to see if the lips are closed. If not, then ask the child to close the lips. If there is no response, the parent closes the child's lips. If this places the child in distress, the parent abandons the effort. If not, the parent gives some suggestions, at least three times, for keeping the tongue on the spot and the lips closed all night, and/or for swallowing correctly all night.

In many cases this "sleep talk" succeeds when other efforts have failed. If there is still no success after another two weeks, the assignment is abandoned. If the use of the elastic is bothersome, have the child continue the assignment without the elastic.

The Goal Chart. We use eight skill areas in the chart provided in the previous step. You might use six, or ten. For the goal chart we combine those areas into four: Resting (tongue and lips), saliva, drinking, and eating. The child has been using the eight-item chart for a couple of weeks. We estimate the mean score, in percents or in letter grades, for the last five or six days for those items that now fit under each of our four new categories, and list that as the present score in the "Today" column. For example, if, during the past six days, the patient averages 65% on chewing, 50% on food gathering, and 80% on food swallowing, the score for "Eating" in the "Today" column would be the mean of those averages, or 65%.

A goal score is written in each of the "Goal" columns. The patient's input is solicited in setting these goals. If the patient would like to try for 100% by the end of the three-week period that follows before the next session, today's score is subtracted from 100%. Since there are five steps to the final goal day, that difference is divided by five, and the score added to today's score and placed in the next goal column. This child's second column, opposite "Eating," would be:

100%, less 65% = 35%, divided by 5 = 7%. Add 7% to today's score of 65%, so the eating goal four days from today would be 72%. The next goal, four days later, would be 79%, then 86%, 93%, and 100%.

Phase Four: Retaining Learned Patterns

Step Ten. If the patient needs more time to reach the 100% goal, or whatever score you deem reasonable, continue using the goal chart for another period of time. If not, or if you think that motivationally it would be wiser to move to another type of assignment, assign the "two assignments per day" chart, with priority given to those areas where the patient's scores are lowest. We fill out this chart only after seeing the chart from the previous step and after having talked with the patient and parents. Each day of the week the child attends primarily to the two items on the chart that correspond to that day. The child, for example, who is doing poorest with saliva handling, might have 20 "quiet suckings," two times a day, each time followed by a mint, or s/he may be assigned to count 100 correct saliva swallows with a response counter on that day. This may be assigned for as many as four different days during the week. Scattered throughout the remaining days would be a variety of assignments encompassing eating, drinking, and rest postures.

This chart is also kept for three weeks, but may be repeated according to your discretion. Much to do is made out of the fact that now the patient has *arrived;* corrected patterns are at or near 100%. Keeping them there is going to be relatively easy, if only awareness is maintained at a reasonable level. This step is the beginning of the retention phase of treatment.

Recheck Visits. Assignments vary according to the needs of the patient. An important step, though, is one wherein the patient begins to make his/her own assignments. The therapist becomes the supervisor of the patient, who very ceremoniously becomes his/her own clinician at this point. After checking to see if patterns appear normal and easy, get feedback from the observer(s) on what they have noticed at meals and while the patient was occupied with something else. Give the patient the two sheets, the one listing functions and assignments, and the other that begins, "YOU ARE NOW YOUR OWN THERAPIST."

The client is to find the two functions s/he is doing the very best with, then the two that probably need the most attention. They are written in the first two columns on the practice sheet. From the assignments col-

NEXT APPOINTMENT _____   Date _____

STEP TEN: Two Assignments Per Day. For the next three weeks, you will do two assignments, activities, or exercises each day of the week, as I have written them in for you. Mark a " $+$ " if you carry out the assignment, and a " $-$ " if you do not.

KEEP THIS CHART WHERE YOU CAN'T HELP BUT SEE IT OFTEN!!!!!

DAY:	ASSIGNMENT or EXERCISE	1st WEEK	2nd WEEK	3rd WEEK
SUNDAY 1.				
2.				
MONDAY 1.				
2.				
TUESDAY 1.				
2.				
WEDNESDAY 1.				
2.				
THURSDAY 1.				
2.				
FRIDAY 1.				
2.				
SATURDAY 1.				
2.				

NEXT APPOINTMENT _____

Date: _____

Recheck Visit #1

YOU ARE NOW YOUR OWN THERAPIST. And I am the therapist's supervisor. But you also continue to be the patient. As the patient you try to do everything right, and mainly keep aware of all the things you have learned. As the therapist you make assignments to yourself according to your needs. You will need to use the sheet that accompanies this one to help you make the right assignments.

WEEK	I DID WELL ON:	I NEED WORK ON:	ASSIGNED:	DAYS OF THE WEEK							
1st	1.	1.	1.								
	2.	2.	2.								
2nd	1.	1.	1.								
	2.	2.	2.								
3rd	1.	1.	1.								
	2.	2.	2.								
4th	1.	1.	1.								
	2.	2.	2.								

umn on the other sheet are selected appropriate assignments for improving the two items chosen as needing work. Write the days of the week in the little boxes at the right of the chart, beginning with tomorrow. As the patient does the daily assignments s/he places a plus in the box for those assignments. One week from today the patient goes through the same procedure, selecting strong and weak areas and choosing assignments for the following week.

FUNCTIONS	*ASSIGNMENTS*
FOOD:	
1. Keep tongue back as food comes toward your mouth.	1. Watch the mirror for certain meals.
2. Take reasonable sized bites.	2. Make your own reminder signs.
3. Chew with your lips closed.	3. Have parents signal you.
4. Do not push your tongue against your front teeth as you chew.	4. Use reminder stickers.
	5. Select certain foods to always eat right.
5. Collect food without a thrust.	
	6. Practice with soft foods.
6. Swallow correctly.	
LIQUID:	
1. Keep your tongue on the spot.	1. Make reminder signs.
2. Whenever possible keep your back teeth together.	2. Mirror practice during certain meals.
3. Do not lick your lips when you're through.	3. Drink everything a sip at a time.
SALIVA:	
1. Move it to the back of your mouth correctly.	1. Count saliva swallows.
	2. Quiet and squeaky sucking.
2. Swallow correctly.	
	3. Sugarless mints.
RESTING POSTURES (Day or Night)	
1. Tongue on spot.	1. Stickers and signs.
2. Lips together	2. Signals from parents and friends.
	3. Upside-down watch.
	4. Carry things in pocket.
	5. Write on hand.
	6. Night swallows and elastics.

At the next recheck visit, the chart is examined. We once again emphasize the responsibility of the patient in making assignments, and unless our trust in the judgment of the patient is very limited, we give the patient some choices concerning activities of the next few weeks.

RECHECK VISIT—#2

Make six assignment cards, each one different. When you begin using each card (usually you will use it for a week), write in pencil the first and last days of the seven-day period. At the end of the seven days, take that card down and put another in its place.

ASSIGNMENT: _____

REMINDER: _____

DATE: _____ TO _____

ASSIGNMENT: _____

REMINDER: _____

DATE: _____ TO _____

ASSIGNMENT: _____

REMINDER: _____

DATE: _____ TO _____

ASSIGNMENT: _____

REMINDER: _____

DATE: _____ TO _____

ASSIGNMENT: _____

REMINDER: _____

DATE: _____ TO _____

ASSIGNMENT: _____

REMINDER: _____

DATE: _____ TO _____

First choice. Continue using the chart in the same manner.

Second choice. Eliminate the first column.

Third choice. Eliminate the first two columns.

Fourth choice. Eliminate the first two columns and the checking off columns for the days of the week, leaving only two assignments to do each day, with no chart to keep.

Fifth choice. Reduce the work to one assignment each day for the week.

Sixth choice: No assignments, just general awareness.

In our experience, the child usually uses good judgment and chooses one of the first three alternatives. If the child chooses to do one assignment each day for a week, a sheet containing six assignment "cards" is given, and the child fills one in each week, and posts it in a conspicuous place. The child may choose to make bigger assignment notices.

If the patient is not to receive orthodontic treatment, future recheck visits are spaced gradually farther apart: six weeks, eight weeks, three months, six months, one year. If orthodontic work is to be done, an instruction sheet is given the patient or parents urging them to call for appointments prior to any new orthodontic steps: before braces are put on; after they have been worn for a couple of weeks; before they are removed; when a retainer is received; before the retainer use is discontinued; and a few weeks afterwards. These recheck visits only take a few minutes. The patient is warned to keep a close watch on relationships between upper and lower front teeth. If, following orthodontic work, the open bite or overjet begins to reappear, s/he is to phone immediately for another appointment. Some orthodontists will have their assistant alert you when a patient you are treating is due for another visit with you. That is a very helpful practice, if you can arrange it.

Summary

We have presented an approach to treatment of tongue thrust that has been successful for us. Research has substantiated that this approach leads to a permanent change in tongue and lip behaviors, and that the elimination of the tongue thrust contributes to a greater retention of orthodontically corrected dental occlusion. We continue to modify it and encourage you to do the same.

To this approach can be applied use of videotaping for motivational, educational, and documentational purposes. When administered to children, the program must be geared to their limitations, and reinforcers must be more immediate and primary. Adults need reinforcement as much as do children, but of a different kind. We warn against attempting to shorten the program, and advise careful attention to the pattern-strengthening and maintaining portions.

We work with parents in the therapy room with us, with few exceptions. We depend on feedback from them throughout the course of therapy, and we communicate with concerned dental specialists often, in writing and by phone and personal contact. Readers are encouraged to individualize the program a great deal. If the entire program is not within the capability of the patient, tailor it to his/her limitations, but give it a try.

CHAPTER 12

SUCKING AND OTHER ORAL HABITS

The profession of orofacial myology originally grew out of a concern solely for tongue thrusting. This pursuit so dominated the attention of clinicians that other harmful oral habits were overlooked, or were disregarded even when observed. However, certain types of these behaviors refused to be ignored. First digit sucking, then mouth breathing, were found to be heading a list of trouble-makers, areas that demanded modification before there could be a reasonable hope of permanent remediation of tongue thrust. A more holistic approach evolved. The present-day orofacial myologist should be prepared to recognize and treat any counterproductive function of the stomatognathic musculature. In a few instances, such treatment may require the wisdom to refer the patient on to an allied specialist. And always, it is hoped, with the insight to avoid implicating the child, rather than the child's problem, as the culprit.

General Considerations

Types of Habits

We may start with the useful list of oral habits that was formulated by E. T. Klein (1952) many years ago. It is reproduced with some deletions:

A. *Intraoral habits*
  1. Thumbsucking
  2. Finger sucking
  3. Tongue sucking
  4. Lip sucking
  5. Cheek sucking
  6. Blanket sucking
  7. Nail biting
  8. Lip biting
  9. Tongue biting

    10. Mouth breathing
  B. *Extraoral habits*
    1. Chin propping
    2. Face leaning on hand
    3. Abnormal pillowing positions
    4. Habitual sleeping on one side of the face

To Klein's list could be added bruxing, lip licking, tongue rubbing, and several others. The first three items on the list, thumb, finger and tongue sucking, will be held in abeyance for a moment, so that we may scrutinize the remaining entries.

Etiologies

The etiology of these oral habits must be traced to infancy or early childhood. For example, a review of etiologies by Gellin (1964) refers to three studies, involving a total of over 3,800 children, which agree in finding about 50 percent of infants and preschool children to be engaged in "extra-nutrition" sucking. If not practiced to excess, nor retained too long, we accept this as a normal aspect of development. The other biting, grinding and leaning habits that distress us are similar in their early inception; despite their damaging effects later, not one is started with evil intent.

Proposed causes of oral habits may be divided into three categories: physiological, emotional, and conditioned learning.

Physiological Causes

Enlarged adenoids, a deviated septum, swollen nasal turbinates, nasal polyps, and other physiological conditions often lead to mouth breathing, an undesirable habit in itself and an important factor in tongue thrust. Faulty occlusion may lead to an habitual open-mouth position and to abnormal chewing habits. Rabuck (1971) explains how premature occlusal contacts can result in bruxism. When normal compensatory movements fail to produce a comfortable closure, reflex mechanisms are activated which result in painful muscle spasm. The patient, in an effort to avoid the pain, moves the jaw from one premature contact to another and frequently develops a bruxing habit.

Emotional Causes

With the exception of mouth-breathing, most oral habits are probably caused and/or perpetuated by emotional disturbances. Any condition or stimulus which upsets a child's sense of security or sense of worth may produce tensions which result in oral habits. This topic is the basis for several books, and will not be dealt with in great depth in this chapter. Examples of such factors are:

1. Excessive parental demands, related to cleanliness, mature behavior, and accepting responsibilities.
2. Inconsistency within behavior of either parent, or between the parents.
3. The birth of a sibling.
4. An abnormally high ratio of negative to positive verbal and nonverbal input from parents.
5. Teasing, criticism, or physical abuse from siblings.
6. Rejection from parents, siblings, or peers.
7. Forced inhibition of normal avenues of expression for anxieties and fears.
8. Repeated or prolonged separation from one or both parents.
9. Frequent moves from one locale to another.

Gorelick (1954) found a statistically significantly greater number of foster children between the ages of 6 and 10, and 11 and 15, to be sucking their thumbs than his private patients. He concluded that the greater incidence was due to the general insecurity found in the foster children.

The mouth, an early and perdurable zone of pleasure, is a natural resource for the child or adult seeking relief from anxiety. Its stimulation with a finger or thumb, tongue, fingernail, blanket, pacifier, or cigarette is a universal tranquilizer.

Learning

Oral habits may be randomly learned behaviors. A cheek may accidentally wedge between the upper and lower canines as the person listens during a conversation. It may relieve some of the uneasiness experienced when one is being looked at very intently. The act of cheek biting is thus reinforced and is likely to be repeated during a subsequent, similar experience. Faulty sleeping postures would seem to be the products of selective reinforcement. This third type of cause is not, of course, separable from the first two. An act may be repeated to bring about a desired physiological or emotional state, either to avoid an unwanted stimulus, or to secure a wanted one.

Influences On Speech

It is obviously difficult to talk while sucking one's thumb, biting one's lower lip, bruxing, chewing on a pencil, or sucking one's tongue. Thus, oral habits may directly inhibit speech production. A less direct relationship between oral habits and speech may be found when the two behaviors are linked to a common cause, such as insecurity or general tensions in the home. The number of interrelationships among these three phenomena, speech problems, oral habits, and emotional problems, is limitless. A few of them are presented in the following paragraphs.

Oral habits may produce malocclusions and physiological alterations which, in turn, may have an effect on speech. Thumbsucking can restrict the width of the arch, making sibilant sounds difficult to produce. A number of case studies are presented by Dunlap and Streicher (1970) to demonstrate that lateral lisps often result from oral habits. One student was found to be keeping his little finger under his tongue persistently and consistently. Another held his index finger under his tongue and held it in place by turning the tongue over to hold it between his teeth. Still another held a lock of hair in her mouth across the top of the tongue. Other cases were described, and the authors proposed a theory based on 1,200 cases. The substance of the theory is as follows.

Vegetative oral movements occur prenatally. Postnatally developed speech functions have to coexist with the earlier appearing, more important responses. An efficient oral mechanism functions in both areas effectively. The introduction into the system of abnormal stimuli, whether in the form of external objects, or of a malpositioned tongue or finger, intereferes with normal development processes, and speech development is hindered. A common problem among these patients is a lateral lisp. At times the lisp is not present while the habit is occurring, but is evident when the finger or foreign object is removed.

The presence of a lateral lisp in children whose palatal clefts were closed surgically at an early age is common. The resultant narrowing of the palatal arch makes it difficult for the tongue to accomplish the lateral linguoalveolar and linguodental seal, so a substitute, defective phoneme is emitted. A narrowing of the arch resulting from oral habits often has the same effect on the sibilants.

Mouth breathing can produce a short, incompetent upper lip, which yields its function in the production of bilabial sounds to the upper incisors.

One-sided habits, such as jaw-leaning, object sucking, and sleeping postures can result in asymmetries which require compensatory movements for speech production. Some persons succeed in their efforts, and others accept approximations that fall short of producing acoustically normal speech.

Oral Habits and Malocclusions

Davidian (1957) asserts that pressures upon the mandible of the fetus can affect the size and shape of the jaw in the adult. The fetus' arm may be placed under the mandible in such a way as to inhibit growth. Insufficient amniotic fluid diminishes the amount of protection afforded the growing structures and may produce abnormalities in the face.

In the young child, persistent, vigorous thumbsucking can tilt the maxillary incisors labially and/or laterally. The mandibular incisors may be tipped lingually as well. Leaning on a fist may inhibit the growth of the mandible.

Bilateral pressure against the mandible produced by leaning both hands at once may, in addition to restricting the forward mandibular growth, result in a bilateral crossbite in the molar regions. Fluhrer (1975) stated that chin leaning can cause the bite to be closed if the pressure is exerted on the underside of the chin. He claimed to have observed a number of skulls in which the anterior half of the mandible was bent upward, revealing the effects of such pressure.

Gellin (1964) contends that thumbsucking has a deleterious effect on children with "poor" bites, but has little or no effect on the "good" bites. Gellin refers to a study by Benjamin (1962) which found definite correlations between malocclusions, in primary and permanent teeth, and thumbsucking.

Warren (1958) observed 100 eleven year olds in Denmark to determine habitual mentalis muscle function, then examined their occlusions. He found that 67 percent of the patients with malocclusions showed a marked mentalis muscle function, whereas in only 7 percent did the muscle remain passive. Among patients whose occlusion was normal, the figures were almost reversed; in only 5 percent was the mentalis muscle active, and in 62 percent, passive. Of the 100, 31 children had active mentalis function; 90 percent of the 31 had malocclusions. Ninety-two percent of the 39 patients with a passive mentalis muscle had no malocclusion.

Posen (1972) studied maximum lip and tongue forces in 135 subjects aged 8 to 18 years, and drew several conclusions regarding tongue and lip functions and malocclusions. One of these related to Class II, Division 2. He felt that he could account for the lingual tipping of central incisors and the labial position of lateral incisors on the basis of excessive force exerted by a strong perioral musculature.

Posen's (1972) findings also implicated lip habits in the development of bimaxillary dentoalveolar protrusions. In this case the habit is an habitual lips-apart posture. Patients with this habit consistently exhibited a low maximum lip pressure. The absence of constant lip pressure on the anterior teeth allows them to migrate forward, and severe periodontal problems may occur at a relatively young age.

Biting or licking the lower lip can cause a forward shifting of the upper incisors. Gingold (1978) attributes open bite to digit sucking, lip sucking, or tongue thrusting; crowding and rotation of anterior teeth to nail biting; overjet to tongue thrust, mouth breathing, lip biting, or lip sucking; incisor crowding or lingual inclination to excessive habitual neck and head postures. Gingold's plea is to eliminate these deleterious habits early as a preventive measures.

Specific Oral Habits

Mouth Breathing

The preeminent nature of open-mouth resting posture entitles it to be considered first. The orofacial myologist may well spend more time dealing with mouth breathing than with all other oral habits combined.

Causes. The causes of mouth breathing were described above. Basically, any impairment of the nasal airway can cause oral respiration. It is not the result of indolence. In some patients, the antigravity muscles in the tongue and jaws have not developed adequately, so that resting tonus is not equal to the task of maintaining normal elevation and allowing effortless lip closure. These patients may continue to breathe nasally even with the mouth ajar; whether or not breath passes through the oral cavity, both dentition and deglutition suffer.

Effects. A comprehensive review of literature on mouth breathing leads any reader to conclude that this habit does not have a friend in the world. No one has anything good to say about mouth breathing. Its causes and effects are quite well known by the lay public. Our discussion will be limited to its relevance to occlusion. Typical beliefs among dentists regarding malocclusions resulting from mouth breathing are summarized by Sood and Verma (1966): (1) The upper incisors are protruded and spaced, due to the loss of the molding effect of the closed lips. (2) The upper arch is narrowed, due to the loss of the molding effect of the tongue. (3) The maxilla becomes V-shaped, due to the contraction of the buccal segments and the protrusion of the anterior teeth.

The observations of Sood and Verma have been tested experimentally by Paul and Nanda (1973), who analyzed dental model of 100 15- to 20-year-old males, equally divided into mouth breathers and nasal breathers. They found significant differences in maxillary arch dimensions and in anterior occlusions between the two groups. Maxillary arch width was less in the mouth breathers. The authors postulated that this lengthening resulted from the contraction of the width of the arch. Although the palatal arches in the mouth breathers appeared to be lower, no significant difference between groups was found.

Both overjet and overbite were more prevalent in the mouth breathing subjects and in both cases the difference was highly significant (P < .001). Seventy-four percent of the mouth breathers had a Class II malocclusion.

The finding of Paul and Nanda (1973) contradicted those of Linder-Aronson and Backstrom (1961), who found no significant differences in overjet, overbite, or arch width between groups of nose breathers and mouth breathers. A later study by Linder-Aronson (1974), however, obtained results essentially in agreement of those of Paul and Nanda. This research dealt primarily with nasal airflow and adenoidal tissue, and found that children with large adenoids had low nasal airflow, as would be expected. Linder-Aronson also found that children with large adenoids held the tongue habitually low in the mouth. Significant relationships were found between large adenoids and a narrow upper arch, crossbite or a tendency to crossbite, and retroclined lower and upper incisors. Linder-Aronson attributed the retroclined upper incisors to be due to the influence of the muscles of the upper lip when the mouth was held open.

Another study by Nanda and colleagues (1972) was conducted on 2,500 2- to 6-year old children in Lucknow, India. Mouth breathing was found in 27.3 percent of the children. Relationships between the presence of tongue thrust and mouth breathing were not reported, but children having both habits were found to have fewer Class I and more Class III occlusions. There was no significant difference between these children and children with no oral habits with respect to the incidence of Class II occlusions.

varying degrees of distoclusion in 85 percent of the children with mouth breathing he studied.

Watson, Warren, and Fischer (1968) found no relationships between the presence of mouth breathing and skeletal classification in 20 orthodontic patients with mouth breathing.

As usual, in studies concerned with human subjects, research findings relating mouth breathing to occlusion are inconsistent. One of the most important reasons for the inconsistency is the subjectiveness involved in labeling a subject as *either* a mouth breather *or* a nose breather. In reality, most people are either *predominantly* one or the other, and methods for determining the *degree* of mouth or nose breathing are subjective, even when considered, and are rarely even considered.

Nevertheless, the preponderance of studies has found significant relationships between mouth breathing and structural characteristics of the oral cavity. Subjects who habitually breathe through the mouth tend to have narrow maxillary arches; crossbites in the molar area; either overjet or retroclined upper and lower incisors; overbite or open bite, and something other than a Class I occlusion. The presence or direction of cause and effect relationships have not been determined, but most writers implicate mouth breathing as a causative factor in oral structural abnormalities.

Treatment. The orofacial myologist should be trained to make a determination, after obtaining a case history and examining the oral cavity, of whether a referral to a medical doctor for further evaluation is necessary. If the clinician concludes the habit may be only functional, s/he should help the patient eliminate the problem as a first step in treatment of tongue thrust. If not, referral to an ear-nose-throat specialist, or an allergist, is necessary. Most patients who habitually mouth breath can now be successfully treated by surgery, medication, or desensitization measures. Either in the absence of the need for medical treatment, or following such treatment, the patient proceeds more effectively to establish nasal breathing with help from a therapist. Specific suggestions for promoting habitual lip closure are given below under "Lip Habits," and in Chapter Eleven.

A number of mechanical devices have been developed to aid in establishing nasal breathing. Some frequently prescribed are: (1) an oral screen; (2) the Andresen appliance; (3) the Bionator; and (4) a chin strap, either the sort that the orthodontist may use for Class III malocclusion, or the type sold commercially as an "anti-snore mask".

The difficulty with these devices comes when the patient *stops* using them. If the "weaning" can be done gradually, chances of success are greater. If their use is discontinued abruptly, a return to old habits frequently occurs. Nearly all our referring dentists use them only as a last resort, when all other approaches have failed, if at all.

Bruxism

Bruxism is usually defined as any nonfunctional grinding or clenching of teeth. "Nonfunctional" refers to the vegetative functions, of course, because bruxism may well be serving some psychological function. The loud noises produced by bruxing during sleep often cannot be reproduced by the same person during waking hours, which is indicative of the great amount of force being applied to the biting and chewing surfaces of the teeth. In other patients, the grinding occurs without any noise, and the patient is unaware that he has the habit.

The incidence of bruxism, as reported by Reding, Rubright, and Zimmerman (1966), is 5.1 percent in the 16- to 36-year age groups, and 15 percent in the 3- to 17-year group.

Causes

The habit is most frequently attributed to psychogenic factors or to local irritating factors or combinations of the two. Psychologically, the bruxer may be expressing anger, hostility, or tension.

An excellent summary of the problem of bruxism is written by Meklas (1971), who lists several *local* factors in bruxing:

1. Discrepancies between centric relation and occlusion.
2. Tipped or otherwise malposed teeth.
3. High restorations.
4. Chronic inflammation of the periodontal membrane.
5. Differences in or uneven eruption of the teeth.
6. Presence of unusually steep cusps.
7. Tight occlusion.
8. Overcarving of restorations.
9. Rough or chipped enamel or restorations.

The bruxing, then, represents an effort of the patient to alleviate pain, discomfort, or pressures caused by dental irregularities, the "self-equilibration" of the occlusion that was noted in an earlier chapter.

More recently, an unusually complete review of bruxing literature is offered by Gallagher (1980), who includes, in addition to the local factors above, such systemic factors as gastrointestinal disturbances, allergies, and endocrine disorders; psychological factors such as aggression and fear; and occupational factors, such as jobs that involve a great deal of stress or precision activity.

Effects

Symptoms of bruxing include: (1) nonmasticatory occlusal wear; (2) soreness and/or sensitiveness of the teeth; (3) loose teeth due to periodontal damage; (4) muscle fatigue, spasm, or pain; (5) unusually strong muscles of mastication; (6) temporomandibular joint pain and/or clicking; (7) cheek, lip, or tongue biting; and (8) headaches.

Treatment

Treatment methods vary, but dentists agree that the first step is to try to determine what factors are contributing to the persistence of the habit. Any remediable local factors should be taken care of first. Various appliances are constructed to preclude harmful dental contacts. Direct instructions to the patient are given, such as resting with the lips together and the teeth slightly apart. As in the treatment of mouth breathing, suggestions for relaxation and proper resting postures of the teeth and tongue may be given while the patient is going to sleep or during sleep. Hypnosis has been found to be helpful with several of our own patients. Some clinicians use "negative practice," wherein the patients grind their teeth several times a day consciously. It is relatively easy to achieve temporary improvement in bruxing patients, but relapses are common. If repeated failure in therapy occurs with a patient, referral for a psychological or psychiatric evaluation is in order. Programs of medication have helped in certain cases.

Malposture of the Lips

Causes

The causes of lip malposture trace directly to lip nonfunction in mouth breathing, and lip malfunction in deglutition, plus a few inborn jaw relationships, such as extreme upper or lower protrusion, which impel the lips in improper directions.

Effects

The lips exercise an optimal molding effect on the upper and lower teeth when they rest together habitually with moderate tonus and avoid involvement in any biting and sucking habits. Normally the lower lip at rest covers the lower one-third or one-fourth of the crowns of the upper incisors. In this position, it has leverage against the free end of the tooth and is most effective.

Most harmful lip habits are those of the lower lip. However, when the upper lip is short, habitual lip closure is difficult and may contribute to functional (chewing, swallowing, speech) and nonfunctional contacts between the lower lip and upper teeth. Instead of exerting a light, constant pressure against the labial surfaces of the anterior maxillary teeth as it normally does, the lower lip pushes against the lingual surfaces, and is often implicated as the principle etiological factor in labially tipped upper incisors.

Treatment

If the anterior malocclusion is so severe as to preclude habituation of lip closure at rest, ask the orthodontist whether it is possible to move the maxillary incisors toward normal positions as either a first step in treatment of the total orthodontic problem, or as a temporary procedure before total treatment procedures are initiated.

In the usual case, we make lip exercises the initial step in tongue-thrust therapy. Integration of such exercises into the program were detailed in Chapter Eleven; some of the procedures that we have found helpful in reposturing lips will be supplied herewith.

Before we start, it is helpful to understand the opposition. Recall the ten "muscles of facial expression" that comprise the facial network. Directly or indirectly, nine of these interconnect at some point with the tenth, the orbicularis oris, and serve as antagonists. In other words, every muscle in the oronasal region of the face tends to open the mouth, with the exception of the orbicularis oris, which alone closes it.

The contraction of a given muscle is more obvious and dramatic than the relaxation of its antagonist, which permits such action. Strengthening oral closure is of definite assistance, but it is doubly effective if it is accompanied by a positive reduction in nonclosure. Malposed lips are characterized not only by a flaccid orbicularis oris but by a customary hypertonicity in the muscles that oppose oral closure. Some authorities have despaired of such hypertonicity, ascribing it to emotional states beyond control. We generally find it accessible to reasonable mechanics.

Hot Salt Water. From near the beginning of this century, authorities have noted the salubrious effect on hypertonicity of properly applied hot salt water. It tends to increase the blood supply, quickly and markedly, to the area it bathes, as it promotes relaxation of vasoconstrictors, thus stimulating circulation. The warmth itself is relaxing, the increased blood supply speeds the modification of tissue, and if the orbicularis oris remains in contraction while other fibers are stretched, our cause is greatly advanced.

Two or three times daily, depending on individual need, the patient should place 4 or 5 ounces of water in a glass, as hot as possible short of oral discomfort, into which a teaspoon of salt is stirred. A mouthful is then forced outward against the cheeks, allowing these structures to relax completely; the water is held in place for a few seconds and then slowly withdrawn into the mouth proper, but with a feeling of still further relaxation rather than by forceful contraction of the buccinator. Continue in this way, alternately forcing the water from buccal to oral space until the cheeks have been fully distended five times.

With a second mouthful, slowly force the water in similar fashion behind the upper lip, gradually building the pressure until the upper lip is rounded out to its fullest extent. The water is transferred from oral to vestibular space an additional five times, after which the lower lip is subjected to the same procedure with a third portion of the water. Thereafter, the remaining water is used in alternate quantities behind the upper and lower lip. A preponderance may be assigned to one lip or the other if the need is apparent, but both lips should receive attention.

When the upper lip has become short and thick, very little water can be forced behind it at first, and the patient should take careful aim at a sink or washbasin, since initial efforts may result in a spray of salt water. Similarly, the mentalis, if overdeveloped in its usual manner, tends to contract with any oral activity, leaving only a small rim of lip that may be expanded. The refractory lip must then be further stretched concurrently by means of other exercises.

The use of hot salt water is a valuable exercise in itself. However, it may well be made a preliminary for any other exercise chosen for the purpose of stretching and relaxing muscle fiber and surrounding tissue.

Forced Air Version. It may be found useful to establish a routine of performing much the same action as outlined above, except that mere breath replaces the water. The patient forces air behind the most rebellious lip, holds it for a moment, and then releases it. This is to be done several times, after which both lips are inflated simultaneously and the breath held in place for the duration of a slow count to 10. This entire procedure should consume only a minute but should be repeated at intervals throughout the day.

Overlap Exercise. This is simply an exaggeration of the action used in applying lipstick. It is a remarkably effective exercise for the short, thick, or everted upper lip.

Immediately after the salt water procedure, the patient is asked to pull his/her upper lip down over his/her upper teeth as far as possible and hold it there. The lower lip is then lapped over the upper lip and stretched upward as far as possible; this maneuver may require digital assistance at first. The lower lip is then contracted firmly against the upper lip and the mandible slowly dropped, thus massaging the upper lip forcefully downward with the lower lip.

This exercise is repeated five times and is to be done at least three or four times daily. On each occasion one additional overlap is attempted until the upper lip is being massaged downward fifteen or twenty times at every practice period. The use of hot salt water should precede this exercise as often as possible, but no opportunity should be lost to practice simply because the water is unavailable.

After three weeks of this exercise as a steady diet, the upper lip often appears to have grown downward a considerable distance. Actually it has not, of course, but the tissues have been redistributed in a gratifying manner.

Bilabial Gymnastics. When lip closure is accomplished only with an obvious strain in the facial network, it often helps to have the patient "mug" repeatedly during the day. S/he should watch a mirror while doing so, and pause occasionally to rest, but 5-minute periods should be devoted to this undertaking.

The basic idea is to contract the orbicularis oris and then, using only facial muscles, distort the mouth into as many contortions as possible. The lips should be pulled as far to the right as possible, and then to the left. They should be worked up and down, then in a clockwise circle as large as possible, then counterclockwise, etc.

When fatigue sets in, the mouth can be opened and stretched laterally with the fingers, first with only index fingers, and then with two fingers, pulling the lips buccally. While doing so the manual straining should not be resisted by the facial muscles; the latter should remain passive.

The grimacing may then be resumed, attempting to stretch every muscle in the face while maintaining a closed mouth.

Cotton Rolls. We tend to lean rather heavily on the use of cotton dental rolls. These are obtainable from almost any dental supply company, and come in assorted sizes, commonly packaged 2000 per carton. It is best to buy the 1-1/2 inch length in both size 2 and size 3 (Fig. 12-1). The latter is often prewrapped in groups of twenty-five within a box, and many pleasant hours can be spent in wrapping the size 2 rolls into similar bundles of twenty-five. This amount allows for the use of one each day for a three-week period, plus the four that fall on the floor and get dirty.

These rolls are used to form a "lip bumper" behind the offending lip. Since they have an outside covering of gauze, which disturbs some patients because dry gauze scrapes against their teeth, they should be dampened before use, rendering them softer and more acceptable. If the taste of disinfectant is unpleasant, the child is allowed to dip them into lemonade or a similar beverage. Also, it is politic to advise pinching the ends before use, creating a tapered terminal less likely to gouge the lip.

Many more size 2 rolls will be used, since most children can manage only this smaller mass. Older patients will require the larger size 3, although sometimes it is prudent to supply a few small ones for use the first three or four days until the larger size can be accommodated more easily. Note that they are of a consistency somewhere between that of the hard and soft tissues; they are sufficiently firm to influence soft tissue alignment but far too yielding to displace teeth or to render anything other than a helpful effect.

Cotton rolls are indicated for grossly everted lips, for the short or retruded upper lip that does not seem to be responsive to other exercises, for the lower lip that has difficulty in surmounting the edge of protruding upper incisors, and for the obviously overdeveloped mentalis muscle that continues to exist as an unsightly knob on the chin below a deeply creased mentolabial sulcus. In the last instance, cotton should be brought into use rather quickly.

Figure 12-1. Cotton dental rolls, sizes 2 and 3.

Here, again, the result will be much more marked and immediate if hot salt water is used prior to insertion of the dental roll. As a typical example, the patient may perform the salt water exercise when a period of calm is imminent, then do the overlap exercise for a time, and then pop a cotton roll behind the indicated lip, keeping it there for an hour. Either lip may be so influenced, although this procedure is used most frequently to stretch and deenergize the mentalis.

In some few cases, cotton rolls may again be called on during the recheck period. When lips continue to flare apart in repose, they may be redirected into contact by inserting a cotton roll in the vestibule before lips are closed.

For the Nonthruster. Lip exercises such as those described above may also be of great help to the orthodontist in the absence of any tongue-thrust problem. Some patients develop abnormal facial posture because of factors unrelated to deglutition, as noted above. The need for restoration of an optimal balance in the facial network may remain after dental treatment.

Also, a fairly sizable group has been referred for swallowing therapy who were found on close examination to be swallowing in a basically normal manner. It has usually been possible to assign an intensive schedule of lip exercises, often including also some strengthening of the antigravity muscles; these cultivated abilities were then transferred as necessary to resting posture and to the act of deglutition, and the entire formal program was avoided.

Oral Screen. For those cited in the preceding paragraph, and for many others undergoing full-blown tongue-thrust therapy, the day arrives when they must establish nasal breathing along with the ability to close their lips firmly without undue strain. This objective may be reached much more quickly and easily if the patient is supplied with a modified oral screen.

In its formal version, the oral shield, or vestibular screen (Fig. 12-2), is carefully built of acrylic from a model of the patient's teeth, as described near the first of Chapter Four. This type of shield is beyond the needs or competence of the clinician. However, a light, flexible creation based on the same idea may serve admirably for present purposes.

Two basic sizes are made, as in Figure 12-3, and stored individually in coin envelopes; one of them can then be cut to fit the patient's requirements. The screen is made of low-density polyethylene plastic 0.030 inch in thickness, the same as that used for lids on many coffee cans and soft margarine cups. This product is difficult to acquire in localities that do not have a well-stocked plastics distributing company. However, many glass and mirror companies are in contact with such distributors and may be willing to put in a special order for you. A minimum quantity is often a 4 by 8 foot sheet, costs very little, and is sufficient to fill your needs for several hundred years. Share it with a friend.

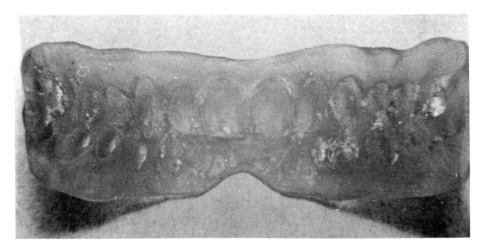

Figure 12-2. Vestibular screen made of acrylic.

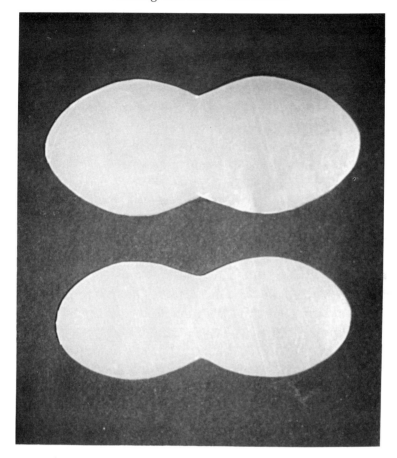

Figure 12-3. Child and adult oral screens of polyethylene.

The notches in the center must be sufficiently deep to avoid contact with the upper and lower labial frena, since the frena may be quite sensitive to pressure. Nevertheless, the overall size of the screen should be adequate to fill the vertical dimensions of the vestibule just short of discomfort.

When it is properly fitted, the patient can occlude both teeth and lips, although lip action is stimulated by the need to maintain the screen in place. It is now impossible to breathe through the mouth, for the inner surface of the lips seal against the plastic, and any attempt at oral breathing strengthens the seal. Therefore *oral screens should never be used in the presence of nasal blockage.* Should temporary blockage occur, the patient has only to touch the screen with his tongue in order to dislodge it. Some shields have been produced with breathing holes cut along the midline; this arrangement defeats the very purpose for which we are striving. The air holes convert instantly to whistles.

It has often been found valuable, after the patient is accustomed to maintaining the screen for an hour or more during the day, to require that it be kept in place during sleep. Many patients who habitually mouth-breath at night are later found to be sleeping with their mouth closed.

In certain orthodontic relapse cases that are caught early, or in some Type I cases in which only incisors are disturbed, it has seemed advisable to suggest that the dentist construct a formal acrylic shield, thus effecting tooth movement at the same time that lips are being closed. The dental version is invariably more effective.

Lip and Cheek Biting

Lip biting and cheek biting are relative infrequent habits, at least to a pathological extent. Biting with sufficient force to displace teeth becomes painful. It is a common reaction to certain types of stress even in the well-ordered mouth, on a periodic basis. Unfortunately, a few patients do persist to the point of habit formation, to the detriment of their teeth.

These are usually among the less difficult habits to displace. Once the causes and effects have been explained to the patient, it remains only to call up to consciousness the instances of habitual occurrence. For the lip biter, this is often possible by keeping the exterior of the lip well coated with a lubricating ointment or paste. The cheek biter often has a shelf of buccal tissue that forms and hardens from incessant nibbling; an oval sheet of the polyethylene plastic used for oral screens can be placed in the buccal vestibule; this allows the shelf to resorb while preventing further cheek biting.

This fibrous buccal ridge of the true cheek biter is readily apparent. Some lateral tongue thrusts are misdiagnosed as cheek biting when in reality the tongue is the guilty agent. These patients display a smooth, normal inner surface of the cheek, even at the site of the posterior open bite.

Lip Licking

Lip licking is vexingly self-perpetuating and self-enlarging. While it is occasionally the result of chronic nervousness, in most cases it is another product of mouth breathing. As described in earlier chapters, breath fanning in and out over the lip (usually the lower lip) parches and chaps the vermillion portion. In an effort to alleviate the resulting discomfort, the tongue coats the lip with saliva. The drying saliva encourages further chapping, but also extends the affected area somewhat below the vermillion line. The excursion of the tongue must necessarily increase accordingly, but never quite catches up to the progressing outer boundary of the red, sore, traumatized tissue.

The primary treatment, of course, is to institute a closed-mouth resting posture. In the interim, as with lip biting, it is helpful to keep the entire outer surface of the lip constantly coated with a cream or unguent.

Lip Sucking

Lip sucking is frequently referred to in dentistry as a "mentalis habit," since contracture of the mentalis muscle in positioning the lower lip may result in overdevelopment and a characteristic "button chin." In orofacial myology, we find that the hypertrophied mentalis is more often the product of maintaining a labioglossal seal during swallowing, and is unrelated to sucking the lip. Be humble in this knowledge.

Lip sucking may represent a transfer when punitive measures have been used to displace a thumb sucking habit. On the other hand, it is often an advanced stage of lip licking. Once the tongue has reached a comfortable limit of protrusion, the alternative of sucking the lip into the mouth is discovered. Once this is established on a habitual level, the evidence is unpleasantly obvious: the lower lip is not only chapped and discolored, but is marked by an angry red semicircle dipping down toward the mentalis.

It is generally impossible to even begin procedures for mouth closure until this habit is disrupted: the instant that the mouth closes, the lower lip darts between the teeth.

Consistent with our therapy for other oral habits, we prefer to avoid the use of appliances as much as possible. Lip sucking drives us to such recourse at times, calling forth an oral screen as described above. Again, the acrylic type, made by the dentist, is to be preferred although the cost is far greater. Our plastic translation is less likely to be retained in the mouth, but is usually satisfactory.

As in the two preceding conditions, it is essential that the lower lip be kept well lubricated on a 24-hour basis; otherwise, the chapped status of the lip creates an intolerable sensation, the screen is yanked out, and lip sucking is resumed.

Nail Biting

On our roster of oral habits, nail biting is more remarkable for its frequency than for its dental influence. Unlike the other addictions with which we deal, that apply pressure *against* the long axis of the tooth, nail biting exerts its force *along* the long axis, a stress that the tooth is specifically designed to withstand.

Incidence. This habit, according to an oft-quoted study by Wechsler

(1931), is found in about 43 percent of adolescents and 25 percent of college students. More recent studies have lowered those figures somewhat.

It is unusual in children under the age of 3, often starts at age 4 or 5, and is found in about one-third of the children from age 6 to the beginning of adolescence. There is a sharp drop in incidence at the age of 16 to about 19 percent. Approximately 10 percent of adults bite their nails, according to Pennington (1945).

Effects. Whereas it may be possible for damage to the gingiva and teeth to result, usually the only physical damage is to the nails. Because it is not a socially accepted habit, its chief damage may be psychological rather than physical in nature. It is quite possible that it is the chronological successor to thumbsucking as a sign of insecurity or nervous tension.

Because of its rather common occurrence in children, the identification of nail biting as a *problem* depends on its effect on the subject. If it is a source of embarrassment to the child or the child's parents, or is causing some physical damage, a logical specialist to whom to refer is the orofacial myologist.

Treatment. A review of treatment procedures is presented by Nunn and Azrin (1976). These include negative practice, operant conditioning and psychotherapy. According to Nunn and Azrin, none of those procedures has been generally effective in eliminating nail biting. The authors' "habit reversal program" consists of the following series of procedures:

1. Clients are trained to heighten their awareness of the nature of their problem.
2. They are then taught activities which are incompatible with nail biting, such as manicuring them, then grasping an object firmly or clenching the fist when tempted to bite the nails.
3. In motivational discussions the undesirable personal and social results of nail biting are reviewed. The support of family members and peers is enlisted to help provide reinforcement for proper behavior.
4. The clients imagine themselves in situations in which they would be likely to engage in nail biting, and demonstrate the use of incompatible activities in those imagined situations.
5. The clients seek out situations where nail biting is likely, and practice the competing activities.
6. The clients inspect their hands and nails nightly and repair any roughness or peeling of the cuticles.

This program was given to thirteen patients, ranging from 11 to 38 years of age, all of whom had bitten their nails for at least eight years. All thirteen eliminated the nail biting within one month after beginning treatment. Only two of the clients demonstrated any relapse during a 16-week period following treatment, and these two clients each had only one incidence of nail biting during that time.

Since we know of no other approach claiming comparable success with nail biters, we recommend that the reader try the Nunn and Azrin procedure. We would add one helpful hint. Aside from its function as a tension-relieving mechanism, one of the most potent inducements causing the habit to persist may be found in the snaggled edges of the nails themselves. These tend to catch as they brush against clothing or other materials. Before the patient even realizes it, the jagged offender is automatically whipped to the mouth, to be gnawed into submission.

The patient should be provided with a goodly supply of emery boards. These should be placed at strategic locations about the house: one in purse or pocket, others near study areas, by the TV chair, on the TV, by the bed — any place that the patient may linger even briefly. With a little time and effort, the habit can be transferred from the teeth to the emery board.

Extraoral Habits

The orofacial myologist should have some understanding of so-called "leaning habits," but may find their occurrence extremely rare. Also, any resultant impairment will usually have been completed by the age children are seen for therapy, obviating any type of treatment other than orthodontic.

Note once again that sporadic pressure does not alter dental development. In order to influence jaw growth or distort the buccal arch form, the unwanted force must be excessive in amount and very long in duration. During sleep, or at rest during the day, we tend to change our posture quite often, and so relieve pressure at any one site.

Should adequate pressure persist, the ensuing deformity is usually seen in the maxillary arch, since the mandible is mobile and gives way under pressure. And except for the rare chin-propping case such as cited by Fluhrer (1975) earlier in this chapter, the distortion is almost invariably unilateral.

As with cheek biting, differential diagnosis may require a second look. it is a learning habit, or a unilateral tongue thrust?

Digit Sucking

Let us state at the onset that we know of no quick and effortless panacea for the management of digit sucking. This is an area that deserves careful and detailed exploration. It has been discussed for many generations in the literature of many professions—psychology, medicine, dentistry, speech pathology, and others. Despite all of these statements and investigations, many clinicians still scurry about seeking for simplistic solutions to the problem. It seems that almost as many others have a ready answer, ranging from the unfounded horror of the Freudians at even raising the question, to the barbed "crib" still welded in place by a few dentists who lack both patience and compassion. Somewhere between must lie reason.

Even this middle ground is fairly well populated, however, and the futility of reviewing all the proposals that have been made soon becomes evident. We have tried a dozen or more procedures, have read dozens of books and articles on the subject, and have managed, one way or another, to achieve the elimination of sucking habits in several hundred patients over the years. Some of the basic concepts thus described, and techniques consistent with these premises, will be outlined. It should be noted that these procedures, none of which are claimed as original, have been remarkably successful when acceptable to parents and patients. They have generally been so accepted, and no damage or untoward aftermath has yet been demonstrated. In fact, definite emotional gains have at times appeared to stem from successful mastery of this situation.

General Orientation

It is strongly felt that the therapist should not rush in and begin tossing techniques about before acquiring some genuine understanding of what is involved and of the guiding principles that serve as the skeleton of the methods being employed. For one thing, success in this venture will depend in large part on the extent to which you are accepted as an authority on the subject. With any given patient who retains a sucking habit there have been lay efforts in wide variety, and over a period of years, before appearing for swallowing therapy. You must offer *professional* guidance.

The one factor that then appears to have the greatest influence on the entire program is the *confidence* of the therapist who attempts to utilize it. A number of people have been observed who did not thoroughly understand why they were making certain suggestions, how these suggestions were expected to operate, or what some of their ramifications might be. Not only are these people prone to error, but they also transmit their own feelings of insecurity to the patient, report an understandable lack of success, but tend to blame the technique rather than their lack of preparation.

Since, therefore, comprehension breeds confidence, and confidence success, let us examine some of the pertinent details. Should your background in this area be superior to ours, you will be able to scan this section with even greater confidence and self-satisfaction.

Etiology in Tongue-Thrust Cases

Sucking is what infants do best. It is the strongest instinct and the most highly polished skill of the neonate. As with tongue thrust itself, sucking habits are so common during the first 2 or 3 years of life that they are accepted as normal behavior. Therefore, our search for causation—again as with tongue thrusting—is not concerned with *why* children suck their thumb, but rather why some children *persist* in this behavior.

It appears that those who continue to engage in digit sucking after they start to school, and those who retain their tongue thrusting pattern of swallowing, are heavily overlapping populations. Once again, there is a frustrating lack of solid research. Nevertheless, from very early on it has been our clinical impression that there is a sharply reduced sucking element in abnormal deglutition. Acceptance of this premise goes far toward explaining the greater incidence of sucking habits among tongue thrusters as compared to nonthrusters. When digit sucking has played a major and fundamental role in early life, helping to satisfy an essential craving, it may be less readily abandoned during the transition period at 3 to 5 years of age when most children drop such habits.

Boundaries of Competence

Before going further, we should explicitly set and define the limits of the population for which these techniques are intended. We seek to entice no one from the psychotherapist's couch; to the contrary, we routinely refer patients to a psychologist when such referral is indicated.

Nor is this program designed for preschool children. A number of clinicians focus on that group, and we wish them well. We have less regard for reports such as that by Modeer et al. (1982), who allege that sucking habits should be extinguished before the child reaches 2 years of age.

The procedures herewith proposed are employed with reasonably "normal," fairly stable young adults and children down to the age of 6 years, who are in average physical health but who swallow improperly. They are not appropriate for children below the age of reason, for the acutely ill, for the emotionally disturbed patient, or the patient who is severely limited mentally. Indeed, we should not be working with such a patient in the area of deglutition, failing some exceptional circumstance.

Emotional Concomitants

We should take a direct look at the emotional effects that are attributed to intervention with digit sucking. Counselors, in discussing sucking habits, often become more emotionally disturbed than the patients over whom they fret. This is not to imply that there are no emotional overtones in this problem. Of course there are, and we should be aware of them. We should not overgeneralize them, however, but rather we should put them in perspective appropriate to their nature and origin.

Scanning the furor over sucking habits in the professional literature confirms the impression that the basic controversy revolves about this one point: the effect of cessation on the personality of the patient.

The continuance of this habit may be more emotionally traumatic than its elimination, in many cases. The child who is subjected to ceaseless harrassment by his family as an indirect reaction to his/her sucking habit, who uses the habit as a weapon against parents, or who dares not contemplate slumber parties or a friend "sleeping over" because s/he lives in dread that his/her habit will be discovered by his/her peers is being warped and harmed in ways other than dental.

As a general observation, we may note the growing accumulation of evidence that human emotions do not actually function in the closed system, which, hypothetical construct though it always was, has been accepted by many workers as verified and gospel. That is, the personality does not necessarily duplicate the seething teakettle, so that if the spout is pinched shut, the lid blows off. It might then be conceivable that one could stop sucking his/her thumb and not turn immediately to masturbation, bed-wetting, or other compensatory behavior. Certainly the use of force and coercion can have a deleterious effect, regardless of the context in which it is applied. The emotional reaction may be to the method, however, rather than to the modification of behavior thus enforced.

Naturally, sucking fills an unconscious emotional need, even if inefficiently, or the child would long since have stopped. In a few cases, true enough, the habit does persist through sheer inertia. Even excluding the latter, the emotional elements affecting these patients are generally of a superficial, tacked-on nature, acquired through mere association over a period of time, rather than the original basis for the habit. For example, an infant falls asleep repeatedly while sucking his thumb— because s/he feels an unsatisfied instinctive need to suck on something— and eventually, through *association* of recurrent common aspects of the situation, sucking becomes a part of rest, so that even with a feeling of fatigue the thumb pops into the mouth. The small child is punished, and while s/he lies in the bedroom sucking the thumb (perhaps for a physiological reason) and ponders the ills of the world, the thumb begins to symbolize refuge from woe, consolation when wounded, and many other things of an emotional nature conjoined by circumstance but less than causative. Thus the patient will now be required to undergo some emotional reorganization. Other ways will need to be found in which to react to some conditions; in other words, the child will be asked to mature a bit. People who do not suck their thumb grow weary, are injured, and meet frustrations; they sleep, weep, curse, and kick, as the occasion demands, with no feeling of an overwhelming need to suck. Our patient will be asked to make similar adjustments. Assistance, guidance, and support will be required in order to do so; we intend to supply them. The patient will be allowed the time and opportunity to work out his/her own solutions to the problem—but only if the desire to do so is sincere. We ask merely the chance to help.

Wiles That Do Not Work

It should not be necessary to insert this section, for the procedures listed below should be obvious in their futility. Yet they are hardy, time-honored specimens and do not easily die, especially after receiving the transfusion supplied by an occasional seeming success. Therefore, let us examine and dispose of them. There should be no possibility that the therapist would not understand the fallacies that entangle and mislead; parents also should be helped to understand that if these devices were truly capable of success, they surely would have demonstrated results before the age at which we see the patient.

Bribery. This is the most common ploy with the older child and may still be in effect when you see the child, or new bribes may be in the planning stage, by either parents or child.

Bribery puts a new value on any bad habit, and most children are capable of recognizing a good thing when they see it. The bribe is usually negotiated like a contract, with specified time periods: "If you don't suck your fingers for one entire month (or two, or more), I'll buy you a new bicycle." The child scrupulously avoids being caught in the sucking act for the required period, gets the bicycle, and rides it around the block, realizing all the way the price s/he has paid for it. Having earned it, s/he parks the bicycle, sits down and sucks the fingers. Any new offers?

When a time period is not set, the bicycle loses luster daily, the task appears more and more hopeless, and the child is driven back to the fingers in despair. The sense of failure is assuaged somewhat by sucking, and the habit gains additional strength.

Rewards in any form of worldly goods are strongly discouraged during the period when we are attempting to eliminate sucking behavior. Cold though the prospect may be, we are asking the child to find reward within—in self-mastery, advanced maturity, increased confidence, a sense of freedom from enslavement to the habit, even pretty teeth. In the daily life of the child, these rewards are less abstract than they appear. They are also permanent.

Punishment. Certainly the oldest, most frequent and earliest answer to sucking is punishment. It probably also comes in the greatest variety of forms, from simple, crude physical bruises, through more subtle refinements of agony and oppression, to outright parental rejection.

If for no other reason, punishment should be avoided because it provides a specific and intensified *need* for sucking. Even at the threat of retaliatory action, the thumb sucker flies to the thumb for solace. The greater the punishment, the stronger and more ingrained becomes the desire to suck. Another vicious circle is complete.

Furthermore, anyone who has worked with stutterers understands the mechanics of *secondary gain*. Although punishment is not pleasant, it does at least provide parental attention, a scarce commodity in the lives of some children. It also provides a convenient cloak for other failures in daily life. Thus, if the child persists in sucking, s/he receives attention and is excused from certain other obligations; s/he may decide the price is right.

We will insist during the following procedures that parents exercise some restraint in this regard. If inadvertent instances of sucking occur, the parents should feel sympathy for the child who is trying to stop, they should express regret that the project is proving difficult, and they should display affection and approval of their child as a person and admiration for his efforts. There is broad scope for responses short of ridicule, shame, or castigation.

A clear distinction should be drawn for parents between punishment for sucking behavior and *discipline.* Nothing that has been said here should be construed to imply that the child's other activities should not be held within some definite boundaries. Otherwise, we may build a greater problem than we destroy. We are definitely not asking that all restraint be removed from this child's life; misbehaving should be penalized.

In this regard, note that it is simply not possible to *love* a child too much; it is readily possible to overindulge him. Overindulgence is not healthy and is certainly not love; it is indifference to the child's future well-being and is thus antithetical to love. When it is easier to tolerate misbehavior than to correct it, the child can only assume, usually correctly, that parents care very little about him/her. The insecurity that results from parental rejection pays the rent for many psychologists and attorneys. In fact, establishing a rigid program of sound discipline may be the kindest service we can render to certain children, as we shall see in a moment.

Gimmicks. It is difficult to find a more suitable title than "gimmicks," for this section covers such a wide range of physical devices purporting to be quick and easy methods of "breaking the habit" but usually proving instead to be something-for-nothing snares for the unwary, the indolent, or the poorly informed.

These devices usually begin with the evil-smelling or vile-tasting solution that is painted on the child's thumb. A tolerance is quickly developed for these products much as learning to eat red-hots or olives, after which the child may enjoy licking off his thumb before sucking it. In quite a similar vein is the application of cayenne pepper to the child's fingers. Once the child has then rubbed sleepy eyes with these fingers, and screamed for an hour thereafter, most parents relent and allow the child to drop off to sleep sucking freshly washed fingers.

This failure often leads to the "old sock" routine: one of father's wool socks is tied over the hand at bedtime, hopefully making the thumb unavailable. If possible, the sock is worked off the hand as the light goes off. If not, or when the sock is retied just short of tourniquet status, a hole is forced in the sock, one way or another, and the precious digit, in one sweeping motion, darts through the hole and into the mouth.

Another favorite is the projecting stick. Preferably, a tongue depressor is cut in half, or any available substitute may be used. It is attached to the thumb, splintwise, with adhesive tape, a goodly portion protruding beyond the end of the thumb. As the thumb approaches the mouth, the stick makes prior contact, bruising the lips, stabbing the palate, or choking the child if it is injected too far before the mouth is closed. The child's allegiance is quickly switched to a different digit, its desirability augmented by the recent oral wound, and the aptitude for sucking grows in stature.

At this point, desperation measures may be brought to bear. Arms are tied to the bed or bound to the body; the child's body may be wrapped tightly in a sheet, arms at sides, in an effective straightjacket. One frantic physician, at the end of his tether, splinted both arms of his avidly sucking, switch-hitting son, and then dared him to find an out; as incredible as it still seems, the boy managed to suck his toe.

As generally employed, we can only include the dental "crib" in this category, as well as the metal bar or other contraption welded into the mouth to make sucking painful or impossible. Dentists deny vehemently that they ever place such devices without explaining the necessity to the child and gaining his/her approval. The number of children who then proceed to break the crib out of their mouth, often at the cost of considerable pain, or bide their time until the crib is removed before resuming their sucking habit, or express other indications that they are submitting to this procedure with less than enthusiasm, leaves some doubt as to the advisability of this approach.

In reviewing the "gimmick" approach, we should be aware of the fact that these devices are basically merely ornamented forms of punishment; beneath the paint and bandages, they remain as crude as a hit in the head.

Corrective Procedures

Prospects for success are brightened considerably if both parents are present for this interview. A small deception is also requisite, in that the patient must have no knowledge of what is discussed at the initial portion of the conference; as far as the child is concerned, you are simply compiling case history material and visiting for a longer time than necessary. The child's turn will come. The session is customarily handled in three segments: parent conference, a private interview with the patient, and a joint gathering with child and parents.

Although it may appear that we are overdoing this program a bit, taking unnecessary precautions and asking an excessive amount of routine from child and parent, it is felt to be justified. No ill can befall the child on this regimen except for a tragic sense of failure if our program falls short. When this procedure is carried out fully, it is adequate to the task of transferring some very deep-seated tendencies to more socially acceptable outlets. Part measures and shortcuts have frequently proved deficient.

Parent Conference

Sufficient information is elicited from parents to form a general opinion of the child's personality and the family situation. The program is then explained as succinctly as possible while still assuring that all vital aspects are understood.

When more than one child in the family has a sucking habit, it is far better to take them one at a time if possible. Rarely has this program proved fully effective as a group project. In the case of twins, or siblings of near the same age who share a room, and given a sturdy mother, it has occasionally been feasible to complete this program on a two-child basis. It is not recommended.

Parental Program. In most cases we will ask that the parents serve as a life raft for the period when the child is floundering in deprivation for the first three or four weeks. They will be asked to increase their overt expression of affection and attention during this time. This is the arrangement of which the child should not be aware, so that it falls as manna from heaven, with no requirement that s/he vie for attention.

Specifically, to satisfy this function, the parents will be asked to (1) make themselves (at least one of them) available to the child at all times during the coming month; (2) provide the child with a few minutes daily when they are absolutely alone and the child is actually held in the parent's arms; and (3) at least twice daily, one or both of them is to openly display attention or affection which is *in addition* to that usually displayed.

The first requirement—that a parent arrange to be at home while the child is at home—may be received as a sentence to jail. When both parents are employed in jobs outside the home, this requirement poses a problem, and some provision for latitude must often be made. In all other situations we are rather rigid on this point; if the welfare of her child is not sufficiently important to justify mother's absence from a tea and a few bridge games, you are probably defeated in any event, and the child may have a legitimate basis for sucking his/her thumb.

The second proviso, that a parent arrange for a private "close time" with the child each day, may also require some planning when there are other children in the home. Assist them in thinking this point through before going on. The father may feel most useful here, in that he may be able to keep the others occupied and distracted while the mother steals away with her "problem child." Alternatively, the children may return from school at different times, younger siblings may go to bed earlier, or there may be some other time that can be found. This problem can always be worked out when there is a willingness to do so.

With younger offspring, the child is truly expected to be held on the parent's lap during this brief period; should the child be larger than the parent, they should at least sit very close, with an arm around the child, or otherwise maintain physical contact.

The additional attention and affection need be nothing dramatic, only consistent, overt, and unsolicited by the child. Not that we oppose something on a grander scale, except when it might startle the child if it is unprecedented. Simple pecks on the cheek are acceptable as a daily diet.

This is the time to ascertain the discipline program in the home and to discuss it on the basis just outlined. You should be sure that the parents understand that they are not being inhibited when a genuine misdeed occurs.

You should also assuage any guilt feelings arising from the prospect of neglecting their other children during this period. If one child became ill, they would feel no compulsion to take the others to the hospital. For the moment, this child has a special need; the others will have their turn as necessary.

We remind them, as we will point out later to the child, that no one is perfect; we expect an occasional failure along the way. If the expectation were for instant and total success, we would probably never succeed.

We also ask that no deadline be set for completion of this project. Once a date is circled on the calendar, it becomes a menace creeping ever closer; the tension thus generated might be sufficient cause for thumb sucking.

Alternate Approach

In rare instances, a child is found who has been so smothered with attention that an additional layer would be suffocating. The program just discussed would then be highly inappropriate. For this child we may fill a desperate need by outlining a schedule for discipline, for both child and parents—often extending to grandparents and aunts and uncles.

It is fortunate that such cases are presented infrequently, for the prognosis is never as good. The type of parent who instigates this situation, or permits it, is not easily dissuaded. Occasionally, however, one sane member of the family has been crying for support, and you may enlist in his or her cause with some effect.

It is generally not possible to speak candidly with these parents, for they cannot accept what you are saying. The obvious results of their overprotection should be commented on as tactfully as possible, and the requirements for good discipline specified.

You can list independent activities that are appropriate to the child's age and suggest that they be made available, for immaturity may be marked. You can note the benefits of assigning daily tasks and of withholding approval until the job is satisfactorily completed.

Telling these parents that they cannot live their child's life is a meaningless abstraction. Instead, you might present them with the ridiculous picture of the parents attempting to insert themselves into the lineup of the game in which their child is supposed to play. They cannot bat in the child's place. The parent's role is that of umpire—the arbiter who calls the fouls and who decides when the child is safe or out. The referee has an essential function; chaos would ensue without him/her.

Just as an umpire never changes a decision, once pronounced, so, too, the parent cannot descend to a child's role and enter into an argument over a decision once made. When a trip to the store is forbidden by the parent, the jaunt is out. Should the child defy instructions, s/he is punished. The parent tells the child to pick up his/her clothes, once, and may give one reminder; if the clothes are not then picked up, the child must suffer the consequences. This is only realistic. When the child learns that s/he need only complain, weep, or procrastinate to reverse the orders, discipline is a myth, the child is emotionally crippled, security is undermined, his/her future is jeopardized, and the parents are remiss.

Although this fact may sail over the parent's head on into space, you may be permitted to point out that the basic ingredient in any discipline program is *consistency:* agreement between parents as to what is right and what is wrong, and an inevitability in the child's day-to-day experience, so that the child finds that the same things are right or wrong tomorrow as yesterday. Parents should decide where they wish to draw the line, but a line should be drawn. The difficult part is to keep the line fairly straight, so that it does not waver unduly.

If you are fortunate enough to put some of these points across, you should proceed with your original project. Observe the parents' reaction closely. You have no right to badger these people. If signs of resentment appear, confess that you are out of your depth and suggest that they discuss their problem with their pediatrician. You are certainly then obligated to contact the physician and present your observations before s/he is attacked by the parents.

Conference With Patient

Once the parents have agreed to accept the role that you have specified, they are left to ruminate on the wisdom of their decision while you get acquainted with the child in a different room. Your chat with the patient is usually brief but is probably the single most decisive and critical facet of this entire program.

The goals of this interview will be to acquaint the child with your posture and province in this matter, to construct a relationship with the child as friends engaging a common problem, and, most vital of all, to sign him/her to a verbal contract to play on your team, eliciting an oral statement of the desire to eliminate the sucking habit. This statement and desire must not be influenced by the physical presence of his/her parents.

There are definite advantages to making this discussion between therapist and patient private. The presence of a parent inhibits certain responses, dilutes and delays the establishment of rapport, and casts some doubt on the validity of the expressions of opinion offered by the child. When the child peers across at a threatening parent and realizes the probable consequence of appearing less than jubilant at the prospect of symbolic amputation of his/her thumb, why should we bother to ask for a reaction?

We attempt first of all to remove some of the stigma with which the child feels branded: we daily see tongue thrusters older than s/he who still suck their thumb. We also define our services as assistance, not coercion; if s/he does not want our help, we will simply stop and go home, and nobody will be mad at anyone.

The therapist is warned never to take lightly, or to take for granted, the response of even the older child when this subject is broached in private. It may seem obvious that the patient would wish to be freed of his/her addiction to sucking; nevertheless, it is a continuing aspect of behavior, a fact of life—s/he *does* suck his/her thumb. S/he does so despite prolonged efforts by those about him/her to effect the demise of the habit. S/he therefore tends to be defensive to the degree that s/he has been opposed by his/her parents. Any act that we thus cherish, and guard against onslaught, we must justify both to ourselves and to others. The naive therapist who approaches a child with the attitude that *of course* the child wishes help with this problem, may be shaken by a flat statement such as: "No, I don't want to quit. I'm just the kind of girl who sucks her thumb." The child thereby effectively restricts the therapists' area of mobility: where do you go from there?

It is safer to scout this terrain rather gingerly at first, approaching it obliquely. Some children readily see the direction in which we are moving and spontaneously voice their desire for help or begin to withdraw in preparation for an attack. Should the child sincerely ask for assistance, the interview is closed, and we get on with the project. In many cases the child will require some guidance in finding his/her way out of the protective shell that s/he has built around his/her habit. Should it develop that in reality the child has no intention whatever of relinquishing the habit, you must abide by your word, stop all efforts to pressure or persuade him/her, and report to the parents that it appears illogical at this time to attempt elimination of sucking, that they must accept this circumstance for the moment, and that as soon as the child experiences a change of attitude, you stand ready to help in every possible way.

The latter situation has arisen infrequently and has often appeared to be a test devised by a skeptical child to prove the sincerity of the therapist. In no case to date has a period of more than two weeks elapsed before the child has called to ask that s/he be allowed to return for help.

When the child verbalizes the desire to stop his/her sucking habit, s/he has, in effect, signed an oral pledge of his/her cooperation and voluntary efforts. S/he should, of course, be warmly praised for this decision, returned to the room where the parents await, and given detailed instructions and explanations covering the balance of the program.

Joint Session

You will start with a much better attitude on the part of all concerned
if you succeed in the immediate transfer of the child's guilt feelings to
some other entity. By pointing out the inaccessible nature of the
subconscious, and then placing the blame for continued sucking on these
subconscious impulses, you have rendered some absolution, freed the
child and parent of part of the load, and to this extent lowered the need
for sucking. We inform the child that all we plan to do is provide
him/her with ways by means of which s/he can command his/her own
subconscious mind to stop bothering him/her with thumb sucking.

Specificity is again the keystone, our unseen ally. These people should
be informed not only as to what to do but *how* to do it. The confidence
gained through precise knowledge increases greatly the probability that
the procedures will be properly put to use. When the family remains
unsure of some of the details, they often omit portions in fear of making
an error.

There are four specific elements that the child should incorporate into
a daily routine. S/he is to (1) suck on a piece of paraffin during periods of
inactivity; (2) sleep with an elastic bandage on his/her arm; (3) have a
piece of adhesive tape on the tastiest finger, and (4) keep score daily on
successes and failures.

Note that each of these items has been used individually as the sole
basis for some regiments, and all have had limited success. When unified
into a total program they have proved effective.

Paraffin. A number of benefits are gained from the use of paraffin.
Even its prospect removes at the outset much of the threat that the child
faces in foregoing his/her thumb; s/he may yet salvage something from
his/her loss. It does satisfy to some degree any residual urge for sucking
activity but places it in a different context: it is acceptable to parents and
therefore requires no defensive underpinning, is not really as soul-
filling as the finger, is more trouble to find and soften, and consequently
is gradually dropped. However, it strikes a pleasant note at the start,
provides an alternative, and thus improves the frame of mind with which
the child approaches the other requirements.

Bubble gum and things of a different consistency from paraffin have
proved to be of little use; the child tends to push them into the buccal
vestibule and suck his thumb. The pliable yet firm effect of just-warm
paraffin makes a critical difference. The wax tubes or bottles filled with

colored liquid and sold at the candy counter are ideal: they require no cutting, taste better, are the correct consistency, and are much more desirable to the child. However, they are frequently unavailable in the summer, when they melt unless refrigerated; they reappear in the fall, with the many varieties shaped into lips, noses, and even giant thumbs, which are a necessary part of Halloween. It is helpful to buy them by the carton when they are in plentiful supply and store them up against the lean periods.

When wax bottles are not to be found, second choice is probably a sheet of utility wax. It is available, hopefully in a bright color, at the dental supply. You may present a 1/4 inch layer to the parent, or to the child if s/he is old enough to wield a knife in safety. Strips 1/2 inch wide can be sliced and then folded over two or three times until the optimum bulk is discovered.

Should it be impractical to supply either of these types of wax, parents may be asked to buy a box of household wax (plain paraffin) at the market and cut it into cubes. Cutting this wax is more difficult, since it must be warmed to make cutting possible but melts if it becomes too warm. Whatever is used, it should be stored at home at room temperature, for it will require further warming in the child's mouth before it reaches a peak of satisfaction.

In rare instances, children have refused the wax as a substitute, disliking the taste or the idea, although cooperating thoroughly with the remainder of the program. When television or boredom has then occasioned some return to the thumb, they become distraught with their failure. Although it does nothing to alleviate the urge to suck and seems a poor second choice, some type of glove has often proved helpful. For girls, a pair of inexpensive but frilly white gloves may be purchased at the variety store and worn during periods of danger. Boys would naturally prefer to wear boxing gloves for television watching, but if they are not at hand, work gloves or cowboy gauntlets may suffice. Should the tempter be a digit on the left hand, a baseball glove may also serve.

Bandage. The parent should be instructed with exactitude as to applying the elastic bandage. The type that is 2 inches wide is most satisfactory, but if a wider one is already in the medicine chest, it may be used. In reality, it is not necessary to pull the bandage tightly enough to decrease circulation in the fingers; simply having it there, sufficiently snug to prevent its falling off, is all that is required. You should insist, however, that the pins holding the bandage in place be of a size that the child can open with one hand. We wish to assign all possible responsibility to the child, and s/he should certainly remove the pins and roll the bandage off his/her arm each morning without unnecessary assistance.

When the child displays no preference between thumbs, using both hands alternately, it is usually necessary to place a bandage on each arm. One elastic bandage may cover a small child's arm with only half its length, leaving a second half for the other arm.

The bandage should extend from 6 or 7 inches above the elbow to the same distance below. It should be pinned at top and bottom, with a third pin laterally inside the elbow.

Adhesive Tape. A roll of standard adhesive tape, ½ inch wide, is preferred. Wider tape may be cut to size if it is already on hand. However, plastic backing of any type should be avoided because it is simply too smooth; it is not startlingly different from the texture of a moist thumb, and its vaunted waterproof character robs it of any potential for stool-pigeon duty. The paper-backed variety sold in a dispensing roll works well and usually provides a ready reflection in the morning of any nocturnal sucking.

It is also essential that this tape not be wrapped around the thumb in a continuous spiral that immobilizes the thumb. One small piece is placed over the nail and a separate one around the thumb above the middle joint. When two fingers are sucked, they must not be bound together, but four pieces of tape are used, one over each nail and one above the first knuckle of each finger. We wish to avoid, as much as possible, any sensation of being bound or disabled.

The child should understand from the outset that the purpose of the bandage and tape is not to *force* him/her to stop sucking, for s/he can still do so. The bandage is elastic so that s/he can bend their arm, scratch their head, fix their pillow, or suck their thumb. These items are merely designed to send messages to their subconscious mind when his/her conscious mind—the part that is really him/her—is lost in sleep and so unable to control her/his thumb.

In a few rare cases in which children may develop a skin rash in response to adhesive tape, they should try the nonallergenic variety. On occasion patients have been asked to obtain a bottle of make-believe nail polish, available at many drugstores and toy stores. Instead of using the tape at night, the girls manicure the nail that they habitually suck, and the boys paint a "pirate flag" on the nail. They then place their hand under their pillow as they fall asleep, the girls thus putting their thumb in a "jewel box" and the boys "locking up the pirate." If sucking occurs during the night, the polish is quickly dissolved and disappears—that is, the jewel is stolen, or the pirate wins this engagement.

Calendar. A blank calendar page can be drawn on a full sheet of paper, then photocopies made in quantity. Again, personal involvement is greatly increased by requiring the child to complete the form by filling in the numbers, etc. Older children are asked to obtain a standard calendar for this purpose. In either case, it further stimulates the child's efforts to realize that the record of his/her progress, being compiled daily on the calendar, will be returned later for the therapists' approval.

Each morning the child should inspect the tape before removing it, and keep score on him/herself for the preceding 24 hours. If, to the best of the child's knowledge, they have not sucked their thumb since yesterday morning, s/he places a check mark (✓) for that day on the calendar. Zeros were once used for this symbol, however, to a schoolchild a zero means total failure, and the implication was harmfully incongruous for some children. When sucking has occurred, the symbol used is an "X," large enough to cover the entire number on the calendar. When ten check marks have been achieved, after the last previous X, the child is instructed to telephone the therapist to report success, receive additional instructions, and make an appointment to begin swallowing therapy. In other words, we are asking that sucking be stopped for only ten days as the criterion for success.

This ten-day requirement would certainly be inadequate to accept as an indication of permanency if no further contact with the patient were planned. However, this is only the beginning. When the child *feels* that he has conquered this problem, the heady wine of success keeps him/her going for a time; after all, s/he succeeded on their own! Once victory is achieved and is duly reported to the therapist, the child will be coming for therapy each week. The first transaction after s/he is seated each week will be a question concerning his/her continued status as an ex-thumb sucker and a restatement of the therapist's gratification with his/her performance; thus it would be a betrayal of her/himself to find it necessary, later, to confess his/her desertion from these ranks.

Final Details. Some children, especially older ones, do receive great reinforcement of their resolve if they are able to telephone the therapist once or twice, usually at bedtime. The therapist should never resent these calls for help; we invite them. In fact, we usually initiate a call after a few days to assure that the program has hit no snags. A few words of approbation, reflecting your approval of the child as a person and your confidence in his/her ability to succeed, can prevent some disheartening failures.

When the child calls to report his/her successful attainment of ten consecutive check marks, s/he receives a lavish drenching in praise; a few kind words should be reserved for the parent also. The child and then the parent are instructed to begin deceleration and descent: the elastic bandage is omitted; after two nights, only the adhesive tape over the nail is used, omitting the other piece, for two additional nights; thereafter, nothing—unless the child wishes to continue sucking wax for a time. The latter contingency arises occasionally and may well be permitted; it falls of its own weight in a week or two, as the child's confidence becomes solidified. However, the bandage and tape must come off on schedule. We do not wish the child to build such a dependency on them that s/he feels incapable of resisting their thumb without them.

Time is usually allowed for this gradual unwrapping to be completed before the child is scheduled to return. S/he should continue to keep score on the calendar, and if any relapse occurs, the child should immediately restore the last item that was removed from their arm or finger. If all goes well, s/he returns bearing a much-used calendar and receives further adulation for his/her efforts, the calendar is enshrined for posterity in a special folder in the therapist's filing cabinet, and the child's momentum is carried over into therapy.

When therapy is not scheduled to begin at once, the period of abstention is increased from ten days to two weeks, and after the complete withdrawal of tape and bandage crutches, the child returns her/his calendar in person, as above, for the same memorial ceremonies.

Tongue Sucking

It is rare indeed when a child with a sucking habit does not quench his craving by "milking" his/her thumb, finger or two fingers. Each item specified in the program outlined above applies with equal uniformity regardless of the digital recourse that the patient elects. Exceptional patients, unfortunately, prefer their tongue. This complicates the problem, since it is difficult to bandage the tongue. Nevertheless, tongue sucking is fully as proficient in destroying recently acquired improvements in deglutition and must as certainly be eliminated before normal function can be expected to prevail.

This problem is often as frustrating for the therapist as for the patient.

It has a repulsive appearance, a damaging consequence, and a resistive nature. Saddest of all, no pat answer can be supplied at this time, and there is no truly effective counterstroke or humane plan for retaliation. Wax becomes a turncoat, enticing the tongue into the characteristic protruded, interdental posture from which the tongue continues to provide comfort while the paraffin merely rides the dorsum, the epitome of futility.

The child should be interviewed prior to the parent conference, if possible, to determine if it is even advisable to attempt elimination of tongue sucking at this time. The gravity of the situation should be explained to the patient: it is not conceivable that a normal pattern of deglutition could even be established much less maintained, as long as this habit persists. Attitudes should be probed and powers of persuasion focused. Only a sincere and abiding desire on the part of the patient will suffice.

Should it appear feasible to begin, the parent program is put into effect as above. As for the child, only the calendar remains of our previous procedures. Even an oral screen has little value: the mandible is dropped, the tongue is protruded, and sucking proceeds. The only alternative appears to be the "antisnore mask" described earlier in this chapter when we examined lip posture. When the child will accept this device as help, rather than coercion, there is some hope of eventual triumph. It is necessary to lower and then pump the mandible in order to suck the tongue in a satisfying manner; this action is inhibited by the chin strap and can be produced only with voluntary effort.

It is necessary to wear the mask during periods of physical inactivity throughout the day, as well as at night. No compensatory sucking sensation can be provided. It is a "cold turkey" approach, and it does not always succeed.

Summary

The terms "orofacial myologist," "oral myofunctional therapist," and "orofacial muscle imbalance" are misleading if the only function of the clinician is to correct a tongue thrust. Early in the development of a private practice, nearly all referrals are for tongue thrust treatment, with a few for elimination of thumbsucking. As the clinician inquires of his/her patient concerning the existence of other oral habits and treats them, referrals begin coming in for a wide variety of habit disorders, even a few having nothing to do with the mouth. If another oral habit is found to exist and is causing harm to teeth, tissues, or emotions, the application of certain principles is recommended:

1. Determine whether any etiological factors still persist. If so, deal with them as effectively as possible.
2. Be alert for signs of insecurity and tensions in the patient. Consider your own training, competencies, and limitations in making decisions to treat or refer to another professional.
3. Explain to the patient and parents the advantages and disadvantages of mechanical devices in eliminating oral habits.
4. Provide a means for the patient to give you regular feedback on progress.
5. Select types and schedules of reinforcement appropriate to the age and needs of the patient.

REFERENCES

Benjamin, L. S. (1962). Nonnutritive sucking and dental malocclusion in deciduous and permanent teeth of the Rhesus monkey. *Child Dev. 33,* 29–35.

Davidian, C. (1957). *A study of oral and learning habits and their physiological effects upon the growing face and dentition.* Unpublished theses, University of Southern California.

Dunlap and Streicher Institute of Speech and Hearing (1970). *A new theory based on oral habits as causal factors—speech development.* Monograph

Fluhrer, A. V. (1975). Some original investigations into pressure habits as etiological factors in dentofacial abnormalities. *Pacific Coast Soc. Orthod.*

Gallagher, S. J. (1980). Diagnosis and treatment of bruxing: A review of the literature. *J. Acad. Gen. Dent., 28,* 62–65.

Gellin, M. E. (1964). Management of patients with deleterious habits. *J. Dent. Child., 31,* 274–283.

Gingold, N. L. (1978). Oral habits and preventive orthodontics. *N. Y. J. Dent., 44,* 148–149.

Gorelick, L. (1954). Thumbsucking in foster children, a comparative study. *N. Y. J. Dent., 20,* 422.

Klein, E. G. (1952). Pressure habits, etiological factors in malocclusion. *Am. J. Orthod., 38,* 569–587.

Linder-Aronson, S. (1974). Effects of adenoidectomy on dentition and nasopharynx. *Am. J. Orthod., 65,* 1–15.

Linder-Aronson, S., and Backstrom, A. (1961). A comparison between mouth and nose breathers with respect to occlusion and facial dimension: a biometric study. *Ont. Revy., 11,* 343–376.

Meklas, J. F. (1971). Bruxism—diagnosis and treatment. *J. Acad. Gen. Dent., 19,* 31–36.

Modeer, T., Odenrick, L., and Linder, A. (1982). Sucking habits and their relation to posterior crossbite in 4-year-old children. *Scand. J. Dent. Res., 90,* 323–328.

Nanda, R. S., Khan, I., and Anana, R. (1972). Effect of oral habits on the occlusion in preschool children. *J. Dent. Child., 37,* 449–452.

Nunn, R. G., and Azrin, N. H. (1976). Eliminating nailbiting by the habit reversal procedure. *Behav. Res. Ther., 14,* 65–67.

Paul, J. L., and Nanda, R. S. (1973). Effect on mouth breathing on dental occlusion. *Angle Orthod., 43,* 201–206.

Pennington, L. A. (1945). Incidence of nailbiting among adults. *Am. J. Psychiat., 102,* 241.

Posen, A. L. (1972). The influence of maximum perioral and tongue force on the incisor teeth. *Angle Orthod., 42,* 285–309.

Prince, R. (1971, February). Bruxism, occlusion and migraine. *Anglo-Cont. Dent. Soc.*

Rabuck, R. H. (1971). *Occlusal imbalance and concomitant oral habits.* Unpublished thesis, University of Texas Dental School.

Reding, G. R., Rubright, W. C., and Zimmerman, S. O. (1966). Incidence of bruxism. *J. Dent. Res., 45,* 1198–1205.

Ricketts, R. M. (1968). Respiratory obstruction syndrome. *Am. J. Orthod., 51,* 495–515.

Schlare, R., and Leeds, D. (1957). The trapped lower lip. *Brit. Dent. J., 102,* 398–403.

Sood, S., and Verma, S. (1966). Mouth habits—mouth breathing. *J. Indian Dent. Assoc., 38,* 132–135.

Warren, E. (1958). Simultaneous occurrence of certain muscle habits and malocclusion. *Am. J. Orthod., 45,* 356–370.

Watson, R. M., Warren, D. W., and Fischer, N. D. (1968). Nasal resistance, skeletal classification and mouth breathing in orthodontic patients. *Am. J. Orthod., 54,* 367–379.

Wechsler, D. (1931). The incidence and significance of nailbiting in children. *Psychoanal. Rev., 18,* 201.

CHAPTER 13

PROFESSIONAL MATTERS

Most of the work done in the area of orofacial myology has transpired during the past four decades. It is a very young field. Great strides have been made during that short period of time. Interestingly, the field was conceived, developed in its early stages, and promoted by orthodontists, who, concerned because of the large numbers of their patients whose malocclusions were recurring after years of braces and retainers, seized upon the tongue as the primary contributor to relapse and devised ways of keeping it away from the front teeth. These orthodontists had enough to do just straightening teeth, so sought out people from fields that dealt with changing habits, specifically oral habits, and landed naturally enough on speech therapists.

The speech therapists, armed with exercises and assignments mainly originated by the orthodontists, added what they knew about tongue and lip functions to the armamentarium, and modified those early programs. In general, orthodontists across the United States found that therapy worked better than did the habit appliances they had developed, and they began to refer patients more frequently. Unfortunately, little or no special training was afforded the therapists, and much therapy was provided that was not effective. Therapists began making claims about the ability of therapy to promote tooth movement. Brief training courses were offered for lucrative fees, and attitudes of dental specialists toward therapists began a downward slide.

In 1974 and 1975, three professional associations, the American Speech and Hearing Association, the American Dental Association, and the American Association of Orthodontists, all adopted a policy statement that questioned the value of oral myofunctional therapy and advised against it until more research was done to establish its efficacy.

In 1976, Hanson quoted from an *Asha Journal* article on the history of the American Speech and Hearing Association:

"One of the primary reasons for the existence of a professional organization is the establishment and maintenance of high standards for that profession. The founders of ASHA were strongly motivated toward this goal. They were acutely conscious of the damage to the reputation of speech pathology caused by the activities of untrained, unethical charlatans who claimed to be "speech correctionists." A sizable number of such individuals operated private clinics in the 1920's where, for exorbitant fees, they guaranteed to cure speech defects by the use of secret methods. There was no formal way in which these persons could be identified by unsuspecting clients. But if a strong Association could be established whose members were known to be ethical, trained professional, this would be a means of separating them from the pretenders. Consequently, a chief concern of the founders was the formulation of the particular requirements to be set for membership."

Hanson then wrote:

"I believe that this same condition exists today in the area of oral myofunctional disorders, and that this is the real reason for the statement from the Joint Committee (on Dentistry and Speech Pathology and Audiology). Too many poorly trained people have been doing an ineffective job of administering therapy. It is not that tongue thrust has been found to not exist, nor that therapy to treat it has been proven unsuccessful."

The reasons given for forming the American Speech and Hearing Association were the same as those of the founders of the then-called International Association of Oral Myology, in 1972. Members of that professional organization have worked diligently, over the years, to promote professional standards, to encourage research, to provide a vehicle for the publication of that research, and to raise the level of training of therapists. A profession is emerging, the members of which are holding their heads high. It is still in embryo stage, but it is a very healthy embryo, whose future appears bright.

Is Orofacial Myology Truly a Profession?

Hanson and Mason (1980) address this question in an article in the *International Journal of Orofacial Myology.* They list, from well-accepted separate books by Lieberman (1956) and Stinnett (1968), characteristics of a true profession, then measure orofacial myology against those standards. Eight characteristics are considered.

1. A profession provides a unique, definite, and essential social service.
2. A profession has a unique body of knowledge.

3. A long period of specialized training is required to perform the services.
4. There should be a broad range of autonomy for the individual practitioners and for the profession as a whole.
5. There should be an emphasis upon the service to be rendered, rather than upon the resultant economic gain.
6. A profession is a self-governing organization.
7. There should be a tried and tested code of ethics.
8. A profession requires continuous in-service growth.

Hanson and Mason (1980) offer the following opinions about the current status of orofacial myology, using these eight characteristics as gauges:

1. Generally, orofacial habits in their total scope are treated only by orofacial myologists.
2. Orofacial myologists, at the present time, cannot lay claim to any specialized vocabulary or information that would meet this requirement. Such a body of knowledge, however, is *emerging*.
3. The period of specialized training of orofacial myologists is variable. An important charge to the IAOM will be to develop accredited training programs in colleges and universities.
4. As an occupational group, orofacial myologists are generally autonomous.
5. Most people in the field are dedicated workers, whose primary concern is the welfare of their patients. There are exceptions.
6. The "International Academy of Orofacial Myology" is a small, but well-intentioned and effective association.
7. The code of ethics upon which the IAOM's code is based, that of the American Speech, Language and Hearing Association, does appear to meet the requirement that it be one that is tried and tested.
8. The IAOM does provide adequate in-service growth to its members.

The writers conclude that orofacial myology, as a field of study, " . . . appears to satisfy some of the characteristics exemplified by Medicine, Dentistry, and Law, while moving in a positive manner toward qualifying in other areas." They then offer the following steps for strengthening the status as a profession:

1. Establish more training in orofacial myology in dental and speech pathology college and university curriculums.
2. Organize and carry out scientifically acceptable and controlled research projects which will determine the efficacy of various forms of orofacial myofunctional therapy.
3. Disseminate information to dental and speech pathology specialists concerning the requirements for certification in the IAOM.
4. Strictly enforce the Code of Ethics of the IAOM.

5. Formalize the in-service training by adopting an objective system of requirements for continuing education for members of the IAOM.
6. Identify and influence more practicing orofacial myologists and more prospective trainees by disseminating information in appropriate professional journals regarding the organization, its journal, its training courses, and its national and regional conferences.
7. Establish the IAOM as a nationally or internationally recognized and registered organization.

Work is still needed on all seven steps.

The International Association of Orofacial Myology

The IAOM is the vehicle best equipped to help professionals complete the above steps. It provides for certification and guides its members to high professional standards through its Code of Ethics.

Certification. Through the IAOM, certification is available to persons who have:

1. completed an undergraduate degree in an appropriate field of study.
2. completed an approved course of training in orofacial myology.
3. passed written and practical examinations prepared by the Committee on Certification of the Association, and
4. assumed the ethical responsibilities designated by the Code of Ethics of the Association.

Code of Ethics

Preamble. This Code of Ethics has been created to encourage the highest standards of integrity and ethical principles. This Code of Ethics has been created by the International Association of Orofacial Myology in hopes of helping raise the standards of practice among those who work with orofacial myofunctional disorders. Any act that is in violation of the spirit and purpose of this Code of Ethics shall be deemed unethical. It is the responsibility of each member to advise the Board of Directors of the I.A.O.M. of instances of violation of the principles incorporated in this code.

Section A.   The ethical responsibilities of the member require that the welfare of the person h/she serves professionally be considered paramount.
1. The member who engages in clinical work must possess appropriate qualifications. This member must be certified by the International

Association of Orofacial Myology or be actively working toward receiving this certification. Student clinicians in a training program may treat patients only under direct supervision of a certified member.

2. The member must not provide services for which h/she has not been adequately trained, and/or licensed (if applicable.)

3. The member shall not accept patients for treatment when benefit cannot reasonably be expected.

4. Whenever "just cause" for continuation of treatment cannot be shown, or benefit cannot be evidenced, the member's responsibility is to release the patient from treatment.

5. Diagnosis and treatment may only be administered on a person-to-person basis, not by telephone, letter, or any other form of impersonal communication. This does not preclude the member from calls and letters for follow-up.

6. Professional ethics prohibit the guarantee of the outcome of any therapeutic procedure(s).

7. Every patient's confidentiality must be respected. No information received professionally shall be revealed to unauthorized persons without authorization from the patient and/or the legal guardian.

8. The member shall use every resource available, including referral to other specialists as needed to provide the best service possible for the patient.

9. The member shall maintain adequate records of professional services rendered.

Section B. The member has certain duties and responsibilities to other professions.

1. The member shall seek professional discussion of theoretical and practical issues toward the colleagues or members of allied professions.

2. The member shall attempt to establish professional relations with members of other professions.

3. The member shall attempt to inform other professionals concerning the services rendered by the International Association of Orofacial Myology.

Section C. The I.A.O.M. member has special responsibilities which are not included in Section A, or Section B.

1. The member may not use any dishonest or misleading information in advertisements. The member must follow the specifications set by the local dental, medical and/or speech and hearing associations.

2. The member may not refuse service on the basis of race, religion, or sex.

3. The member shall be active in educating the public regarding the profession of Orofacial Myology.

4. The member shall be willing to share his/her knowledge, new

research and therapeutic techniques with other members of the I.A.O.M.

5. The member shall not use the name of the International Association of Orofacial Myology to promote classes or products or any other profit making venture. This does not preclude the member to announce membership status or offices held in the I.A.O.M. in a vitae or resume.

6. Members shall not accept third party payment insurance plans in excess of fees for service or for fees of services not rendered.

7. The member may not participate in activities which constitute a conflict of professional interest.

8. The member shall present products s/he has developed to his/her colleagues in a manner consistent with high professional standards.

9. The member shall cooperate fully with the ethical inquiries into matters of professional conduct related to the Code of Ethics.

Specific procedures for joining the Association and for achieving certification are provided in the Association Handbook, issued annually.

Validity of the Profession

The official statement of the Joint Committee on Dentistry and Speech Pathology, issued in 1974, was adopted as the official position of the American Speech and Hearing Association, the American Association of Orthodontics, and the American Dental Association. It reads as follows:

Review of data from studies published to date has convinced the Committee that neither the validity of the diagnostic label tongue thrust, nor the contention that myofunctional therapy produces significant consistent changes in oral form or function has been documented adequately. There is insufficient scientific evidence to permit differentiation between normal and abnormal or deviant patterns of deglutition, particularly as such patterns might relate to occlusion and speech. There is unsatisfactory evidence to support the belief that any patterns of movements defined as tongue thrust by any criteria suggested to date should be considered abnormal, detrimental, or representative of a syndrome. The few suitably controlled studies that have incorporated valid and reliable diagnostic criteria and appropriate quantitative assessment of therapy have demonstrated no effects on patterns of deglutition or oral structure. This research is needed to establish the validity of tongue thrust as a clinical entity.

In view of the above considerations and despite our recognition that some dentists call upon speech pathologists to provide myo-functional therapy, at this time, there is no acceptable evidence to support claims of significant, stable, long term changes in the functional patterns of deglutition and significant, consistent alterations in oral form. Consequently, the Committee urges increased research efforts, but cannot recommend that speech pathologists engage in clinical management procedures with the intent of altering functional patterns of deglutition.

If the intention of the Committee was truly to encourage research, the Statement was misguided. Nothing discouraged research more than to place therapy for tongue thrust into a questionable light. Within the profession of speech pathology, and within any profession that deals with changing human behavior, clinical work *always* precedes research to test the efficacy of that clinical work. In 1974, almost no research, controlled or otherwise, had been published in journals of the American Speech and Hearing Association, attesting to the validity of therapy for *any* disorders treated by speech pathologists: articulation, language, stuttering, voice, aphasia, cleft palate, etc. Hanson wrote a reply to the Statement in 1976, and, later, in a letter to the Resolutions Committee of ASHA, summarized the main points of the 1976 article:

I have excerpted statements from the Joint Committee Statement, and replied to those statements. I list each portion of the larger Statement and follow it with my reply.

1. The validity of the diagnostic label of tongue thrust is questionable.

ANSWER: Several independent incidence studies have obtained strikingly similar results regarding the incidence of tongue thrust at various ages. It is a behavior that is consistently identifiable, among separate studies, and among judges within research projects.

2. The contention that myofunctional therapy produces significant consistent changes in oral form or function has not been documented adequately.

ANSWER: (1) The purpose of therapy, as practiced by the great majority of clinicians, is not to bring about changes in oral form. We neither aspire to that goal, nor make claims to those results.

(2) Several studies, some of which are reprinted and enclosed, have definitively demonstrated changes in function as a result of therapy.

3. There is insufficient scientific evidence to permit differentiation between normal and abnormal patterns of deglutition, particularly as such patterns might relate to occlusion and/or speech.

ANSWER: (1) Reliability coefficients of 0.90 and higher are consistently reported by trained observers on independent judgments of the same swallows. (2) The type of tongue behavior during swallowing consistently reflects the type of malocclusion. Several studies by orthodontists have found strong relationships between type of occlusion and tongue behavior during swallowing.

(3) Several studies have found significant relationships between tongue thrust and speech defects, some regarding the coexistence of the two problems in given children, and others regarding improvements in the one being accompanied by improvements in the other.

4. There is unsatisfactory evidence to support the belief that any patterns of movements defined as tongue thrust by any criteria suggested to date should be considered abnormal, detrimental, or representative of a syndrome.

ANSWER: (1) If "normalcy" is that which more than 50% of the population does, tongue thrust becomes abnormal, according to several incidence studies, after the age of five, and gets progressively more abnormal during the next few years.

(2) It is certain that more research is needed to determine whether harmful effects result from tongue thrusting. There is considerable clinical evidence that this is the case.

(3) Tongue Thrust is not a syndrome. We who administer the treatment have recognized for many years that it is a behavior, involving the pushing and/or resting of the tongue against the anterior teeth.

5. The few suitably controlled studies that have incorporated valid and reliable diagnostic criteria and appropriate quantitative assessments of therapy have demonstrated no effects on patterns of deglutition or oral structure.

ANSWER: I vehemently challenge that statement! I know the literature on tongue thrust, and I know of no studies that fit the above description and find those results. Certainly studies by Case and by Overstake are well-controlled, well carried out, and demonstrate definite changes in swallowing patterns. More recently, a study by Christensen and Hanson, soon to be published in JSHD, demonstrated definite modification of swallowing behavior in six-year-olds, along with facilitation of lips correction, as a result of therapy for tongue thrust. Several studies done five years post-treatment have demonstrated retention of corrected occlusion (orthodontically corrected). The entire field of speech-language pathology was, until the past two or three years, embarrassingly devoid of long-term research to measure retention of corrected speech patterns.

Hanson urged the ASHA committee to consider publishing a revision of that 1974 statement, encouraging continued research efforts without casting an unfavorable light on the clinical work associated with tongue thrust. The request was denied.

Since that letter was written, well over a decade ago, the limited research that has been done has supported the position that tongue thrust is a clinical entity, and that therapy for it is effective in changing tongue and lip habits and in assisting in the retention of orthodontically corrected occlusion. The Andrianopoulos-Hanson study (1987), reviewed in Chapter One, demonstrated the effectiveness of therapy for tongue thrust in maintaining corrected occlusion in patients who had had Class II, division 1, malocclusions prior to orthodontic work.

Unfortunately, the stigma associated with therapy for tongue thrust imposed by the 1974 Joint Committee Statement has settled over the profession like poisonous smog. Steps must be taken to reinstate the providing of therapy for oral myofunctional disorders as a legitimate practice. The evidence must be presented to the appropriate committees of the professional associations that adopted the Statement in the 70's in a manner that will bring forth revised position statements from those organizations. The survival and growth of the profession of orofacial myology, notwithstanding professional pressures directed against it, attest to the confidence thousands of orthodontists and other dental specialists place in its ability to help them to normalize the dentition of their patients.

Training Orofacial Myologists

Very few universities and colleges offer courses in orofacial myology. Speech-language pathologists and dental hygienists, who comprise the greatest body of orofacial myologists, must, for the most part, seek training through private courses offered by practitioners. This unhealthy state has changed very little in thirty years. Ideally, a college or university offering degrees in orthodontics, dental hygiene, and speech-language pathology and audiology, would begin to provide the few special courses, plus clinical experience, necessary to prepare students for certification as orofacial myologists. A dedicated effort on the part of the IAOM should be exerted toward this goal. In the meantime, persons interested in preparing themselves to provide such therapy would do well to take at least two private training courses from different instructors, then observe therapy administered by at least two different persons experienced in the field. Information on courses can be obtained by writing the International Association of Orofacial Myology.

Needed Research

Many unanswered, or only partially answered, questions need research
attention. Many investigations require the joint participation of dentists
and orofacial myologists. Some important questions are:

1. Do malocclusions that persist into and throughout adulthood
   really result in periodontal disease and subsequent early loss of
   teeth?
2. To what extent, or in what percent of patients, does therapy for
   tongue thrust result in diminished malocclusions, without orthodontic treatment?
3. Can therapy limited to resting postures of tongue and lips effectively
   eliminate tongue thrust?
4. What is the relative effectiveness of various approaches to the
   treatment of tongue thrust, in correcting the deviant patterns, and
   in stabilizing the dentition?
5. What are the mean and range of pressures of tongue and of lips
   against the dentition, at rest, and during various functions, in
   persons with tongue thrust as compared with those of persons with
   normal tongue and lip habits?
6. Can upper lips be lengthened through exercise? What kinds of
   exercises, and how much lengthening?
7. In what ways are people with tongue thrust different from those
   with normal tongue patterns, with respect to sensory and fine
   motor skills involving the tongue?
8. What are the effects on developing dentition of arresting digit-
   sucking habits at various ages?
9. How effective is preventive therapy for tongue thrust in children
   ages four and five years?
10. Do tongue-resting postures correlate significantly with the native
    spoken language? For example, do speakers of Spanish, wherein
    several consonants are produced with the tongue against the front
    upper teeth, tend to rest the tongue against the teeth more than do
    speakers of English?
11. How do long-term orthodontic results in people treated with
    removable European-type appliances, and who receive no therapy
    for tongue thrust, compare with those for people treated with
    therapy and conventional full-banded appliances?
12. On what bases do orthodontists who refer some of their tongue-

thrusting patients for therapy, and do not refer others, make the selection?

13. Is therapy for tongue thrust more effective when it precedes, accompanies, or follows orthodontic treatment?

Insurance Coverage

Many children and adults who need therapy for oral myofunctional disorders do not receive it because their insurance company will not cover the fee. At the present time, the overwhelming majority of insurance companies pay for speech and related therapy only when the problem is medical in nature, e.g., stroke-related, or accident-caused. Some therapists have better success than others in convincing insurance companies they should pay for therapy for tongue thrust. An excellent article on this topic is Benkert's (1986). Benkert first describes differences among various types of insurance plans: major medical, HMO's (Health Maintenance Organizations), Medicare/Medicaid, and Dental plans. She outlines the responsibilities, in securing coverage, of the patient, the provider (therapist in this case), and the insurance carrier. Benkert gives helpful suggestions regarding submission procedures.

1. **Preparing Insurance Forms.** A sample form is provided at the end of the Benkert article. It is important to complete the form in a professional manner. If space provided for information is inadequate, attach addendums to the application form. Typewritten entries are preferred over handwritten ones.

2. **Physician's Referral Letter.** Whenever possible, include with the form a letter from a physician or dentist describing the need for therapy, as a part of the total treatment for the patient. Suggested wording is:

> (Patient's full name) has been referred to (provider's name) for orofacial myology. These services are unavailable within our offices. We therefore request (patient's name) be seen by (provider's name), an orofacial myologist.

3. **Coding.** Some policies pay for therapy under "dental" coverage, and others under "medical." Have the patient find out which is more likely with the policy s/he has, and submit the appropriate code numbers for the evaluation and for the treatment. Dental code listings are available from the American Dental Association, 211 E. Chicago Avenue, Chicago, Illinois 60611. Procedure codes for use in submitting medical claims are

available from the American Medical Association, OP 341/6, P. O. Box 10946, Chicago, Illinois 60610. Request the latest edition of the current procedural Code Book. Benkert suggests specific code numbers and terminology. The reader is referred to her article for details.

4. **Responses.** Suggestions are given for follow-up procedures when the insurance company rejects the claim. Appeals by patients or groups of patients, especially when they are part of a large corporation that has insurance contracts with that company, are sometimes fruitful. Speak to "higher-ups" in the company. If your patient is the child of the corporation president or vice president, you have a good chance of succeeding.

Outlook for the Future

The current growth of the profession, positive research findings regarding the validity of therapy for oral myofunctional disorders, the lack of advances in alternative approaches to the elimination of tongue thrust, all signal a very positive future for the profession of orofacial myology. A significant roadblock will be removed when dental and speech associations give official recognition to the profession. A great deal also depends on the efforts of individuals and groups, such as the IAOM, in lobbying for better insurance coverage for treatment.

Certain directions in treatment seem likely:

1. Increased attention to preventive therapy: elimination of digit-sucking earlier than five or six years of age; facilitation of nose-breathing in very young children; providing of rest-posture therapy for four and five-year-olds with minor, but developing, malocclusions.

2. Greater emphasis on cosmetic and psychosocial effects of corrected occlusion and proper postural, eating, and speech habits. Case (1982) writes about the effect on people's judgments of other people based on the appearance of the face and on speech. People with unpleasant-appearing faces tend to be judged as being less intelligent and less desirable as social companions. Persons with severe anterior malocclusions tend to develop a faulty self-image.

3. As more and more adults in their thirties and forties receive orthodontic treatment, therapy for adults will become more commonplace.

4. Plainfield (1977) describes the value of myofunctional therapy in assisting patients with dentures to keep them in place. More treatment may be directed to this population in the future.

5. The scope of the practice of orofacial myologists is likely to broaden,

to include more patients with oral habits other than tongue thrust, and even to the area of dysphagia.

6. More formalized training programs should emerge in colleges and universities.

The profession of orofacial myology survived the rigors of birth, enjoyed a pleasant childhood, passed through a stormy adolescence, and is emerging into a personable, competent adult. The adult years should prove to be the best of all.

REFERENCES

Andrianopoulos, M., and Hanson, M. (1987). Tongue thrust and the stability of overjet correction. *The Angle Orthodontist, 57,* 121–135.

Benkert, K. (1986). Insurance: The name of the game. *Intern. J. of Orofacial Myology, 12,* 11–20.

Case, J. (1982). Myofunctional therapy and facial cosmesis: Position paper. *Intern. J. of Orofacial Myology, 8,* 10–12.

Hanson, M. L. (1976). The joint committee's statement on myofunctional therapy — pros and cons. *Intern. J. of Orofacial Myology, 2,* 13–19.

Hanson, M. L., and Mason, R. M. (1980). The nature of orofacial myology as a profession. *Intern. J. of Orofacial Myology, 6,* 4–6.

Lieberman, Ph. (1956). *Education as a profession.* Englewood Cliffs, NJ: Prentice-Hall.

Plainfield, S. (1977-August). Myofunctional therapy for complete denture patients. *J. Prosthet. Dent., 38*(2), 131–137.

Stinnett, T. M. (1968). *Professional problems of teachers* (3rd ed.). London: Macmillan.

AUTHOR INDEX

A

Adran, G. M., 14, 196
Andrianopoulos, M., 3, 21, 22, 23, 264, 265, 357
Angle, E. H., 14, 63, 70
Arlt, P. D., 40
Azrin, N. H., 326, 327

B

Backstrom, A., 313
Ballard, C. F., 13, 199
Barrett, R. H., 12, 18, 229, 230, 265
Benjamin, L. S. (1962), 311
Benkert, K., 359
Bernard, C. I. P., 23
Best, C. H., 6
Bijlstra, K. D., 196
Bloomer, H. H., 45
Bond, E. K., 13, 199
Bosma, J. E. (Ed.), 201
Breinholt, V., 18, 20
Brodie, A. G., 17

C

Case, J. L., 265, 360
Christensen, M., 208, 265
Christofferson, S., 265
Cleall, J. G., 11, 225, 228
Cohen, M. S., 21, 22, 196, 198, 199, 204, 245
Collins, T. A., 226
Cooper, J. S., 265
Crowder, H. M., 229

D

Davidian, C., 311
Dunlap and Streicher Institute, 310

E

Edger, R., 244
Elliason, G–B, 266
Eversaul, G. A., 241

F

Fairbanks, G., 45
Falk, M. L., 242, 243
Fluhrer, A. V., 327

G

Gallagher, S. J., 316
Gellin, M. E., 308, 311
Gingold, N. L., 312
Goda, S., 234, 235
Goodban, M. S., 40
Goodhart, G. J., 241
Gorelick, L., 309
Graber, T. M., 90, 92, 93, 227
Groce, G., 22
Gwynne-Evans, E., 13

H

Hanson, M. L., 3, 12, 18, 21, 22, 23, 196, 198, 199, 204, 208, 245, 264, 265, 294, 349, 350, 351, 357
Hanson, T. E., 21
Harden, J., 265
Harrington, R., 18, 20
Harvold, E. P., 23
Herman, E., 188, 189
Holik, F., 199

I

Ingervall, B., 266

363

SUBJECT INDEX